英伦风格室内设计

英伦风格室内设计

［澳］汉娜·詹金斯 编著

齐梦涵 译

广西师范大学出版社
·桂林·

images Publishing

图书在版编目(CIP)数据

英伦风格室内设计 / (澳)汉娜·詹金斯编著;齐梦涵译. —桂林:广西师范大学出版社,2018.5
ISBN 978 - 7 - 5598 - 0723 - 6

Ⅰ. ①英… Ⅱ. ①汉… ②齐… Ⅲ. ①住宅-室内装饰设计 Ⅳ. ①TU241

中国版本图书馆 CIP 数据核字(2018)第 052077 号

出 品 人:刘广汉
责任编辑:肖 莉
助理编辑:齐梦涵
版式设计:吴 迪

广西师范大学出版社出版发行

(广西桂林市五里店路9号　邮政编码:541004)
(网址:http://www.bbtpress.com)

出版人:张艺兵
全国新华书店经销
销售热线:021 - 65200318　021 - 31260822 - 898
恒美印务(广州)有限公司印刷
(广州市南沙区环市大道南路334号　邮政编码:511458)
开本:787mm × 1 092 mm　　1/12
印张:21 $\frac{2}{3}$　　字数:30 千字
2018 年 5 月第 1 版　　2018 年 5 月第 1 次印刷
定价:268.00 元

如发现印装质量问题,影响阅读,请与印刷单位联系调换。

目录

8	前言 作者：玛姬·托伊
10	南街142号 桑迪·伦德尔建筑事务所
16	安斯蒂·普拉姆住宅工作室 科平·多克雷建筑事务所
22	阿耳特弥斯谷仓 奇利斯·埃文斯+凯尔建筑事务所
28	奥斯丁住宅 亚当·尼布建筑事务所
34	回水别墅 第五站台建筑事务所
42	BALLYMAGARRY公路住宅 2020建筑事务所
48	贝克特别墅 亚当·尼布建筑事务所
54	布拉肯伯里住宅 尼尔·达什艾柯建筑事务所
60	庭院住宅 达拉斯-皮尔斯-坤脱罗事务所
66	拖拽别墅 阿什顿·波特建筑事务所
72	逃至屋顶住宅 小工作室设计公司
78	折叠住宅 外币兑换所建筑事务所
84	画廊住宅 尼尔·达什艾柯建筑事务所
92	哈特罗别墅 电流建筑事务所
98	黑兹尔巷马厩住宅 都柏林设计工作室
104	19号住宅 Jestico + Whiles建筑事务所
110	艺术爱好者住宅 shedkm建筑事务所
116	痕迹之屋 鹤田建筑事务所
122	赫德尔住宅 亚当·尼布建筑事务所
128	无限住宅 迷幻设计公司
134	诺尔住宅 奇利斯·埃文斯+凯尔建筑事务所
140	小堡别墅 哈德森建筑事务所
146	狮门花园 3S建筑设计公司
152	MAP住宅 SAM建筑事务所
158	橡树巷别墅 凯松·卡斯尔建筑事务所
166	奥克兰斯住宅 布莱德斯通建筑事务所
172	旧水塔住宅 格雷斯福德建筑事务所
178	一上二下住宅 麦卡洛·姆尔文建筑事务所
184	页面巷住宅 柯克伍德·麦卡锡
190	佩妮克罗夫特别墅 内皮尔·克拉克建筑事务所
198	探寻别墅 电流建筑事务所
204	玫瑰园住宅 特里亚那·斯塔克建筑事务所
210	场景住宅 场景建筑事务所
216	斯普林菲尔德农场 安德鲁·伍德公司和Designscape建筑事务所
222	圣约翰-赖伊住宅 玛尔塔·诺威卡
230	茅草小屋 re|form建筑与城市规划公司
236	泰瑞特·卡恩别墅 布莱德斯通建筑事务所
242	温克利作坊住宅 柯克伍德·麦卡锡
250	啄木鸟别墅 电流建筑事务所
256	建筑事务所索引

前言

玛姬·托伊

想象自己在一个平静轻松的避风港,一个舒缓宁静的地方,这里也许是五光十色的,也许是宝石装饰的,还充满了家庭的回忆。再试着想象一下你对家的感觉,它是什么样的?是复杂的、开放的,还是简单的?无论它是什么样子的,它都是专门为你设计的,即展现了你选择怎样与这个世界互动。这里是通过住宅设计实现的乌托邦,在这本精致的作品集中,你会看到新建的住宅,也会看到那些改建的住宅。

设计住宅房屋可以说是最难令所有人都满意的挑战,因为这是一个非常私人化的场所,设计师必须非常了解客户的需求,还要能够解释并加强这些需求以便将其转化为设计。建筑史中充满了无法满足客户需求的错误案例,但是也有许许多多的设计师获得了这种完美的平衡。

一些住宅正是应这种极高的要求而建成的。加斯顿·巴什拉德认为"房屋是一种工具,人们用它来面对宇宙",这句话简直就像是从康斯坦丁·梅尔尼科夫的嘴里说出来的,他很可能是在自己被建筑史行业拒之门外后说出这句话的,所以从那以后,他只能在自己家里释放他的创作热情。而这种创作的结果便是那座他的天才创意与宝石的优美曲线相结合的自宅。

维克多·霍塔下定决心创造出一种用钢铁的材质表现出的美感,这种风格后来成为他的作品的显著特征。他在布鲁塞尔的房子仅用正立面就能吸引住你的注意力,让人想对其后面的场所一探究竟。一旦进入其中,你便会看到那些在大楼中旋转的蜿蜒的表现形式,他,它们毫不费力地将来访者引入房屋的各个角落。"我的房子就是我的避难所,它对我来说不是一个方便的去处,而是一个能够寄托我的情感的地方。"路易斯·巴拉干这样说,他把清晰的线条和明亮的色彩混合在一起,从而为他的客户提供了许许多多这样的、能够满足他们需求的环境,他尤其擅长用水流塑造环境,其灵性的设计仿佛魔法一般。当面对过于陡峭的山坡时,许多人都会认为那里无法成为建筑的选址，约翰·劳特纳则不这么认为，他想出了一种巧妙的解决方法，可以帮助他的客户在那样的地方建造房屋。他建造了被称为"臭氧层"的建筑，这座建筑由中央的圆柱支撑，仿佛可以旋转起来，只在一侧轻柔地与斜坡相连接，以便人们可以进入其中。

有时，那些我们认为无法战胜的挑战正是激发创造性灵感的触发物。

如今的现代主义是民主的、不拘一格的。它不拘泥于规则和设计约束，对各种色彩、线条和纹理也都敞开大门。本书中的每一个案例都展示了设计师在排列布局纹理、形状和体积方面的高超技巧，以及由此产生的一系列时尚、精巧、而又令人放松的灵性空间。除了能够处理好设计师和客户之间的沟通，本书介绍的建筑师们还具备能够媲美其优秀设计师前辈们的远见卓识和高超技巧。例如，埃德温·路特恩斯爵士本着坚持时代要求的精神，熟练地掌握了使传统适应现代的技法。对于在现代建筑领域遵崇历史传承的不懈追求也反映在DOM住宿与生活的创始人玛尔塔·诺威卡的作品中，其为20世纪50年代建造的圣约翰救护站带来了21世纪的新功能，同时精心保留了该建筑的历史本质。2020建筑事务所的BALLYMAGARRY公路住宅受到了高度的赞誉，它既保留了历史悠久的爱尔兰乡土建筑的特色，也完美地融合了现代的设计理念。

大卫·奇普菲尔德曾说过："你不能通过填补缺失的部分修复《最后的晚餐》，你要做的是保存它，接纳这些不知何故幸免于难的材料。"

这些住宅对于形式和材料的实验是开放的，但是它们依然能够提供适当程度的舒适，还包含一定的对传统和现代的建筑理念的富有戏剧效果的表达，同时要能够利用当地的气候条件。本书中的房屋设计包含室内空间及其临近的用于娱乐的庭院，设计目的则是获得适当的自然光线，并把极端天气屏蔽在房屋之外，同时也要关注房屋的可持续性及客户对于建造成本的负担能力。

所以，本书中的建筑师们将现在和未来的理念放在心上，肩负着设计完美家园的重任。建筑师、客户和规划师之间为实现这些相互冲突的愿望进行着旷日持久的斗争，它对任何一方来说都是胜少败多的艰难战役。但是，这本书的每一页都向我们展示着建筑师们是怎样把"家"这个现代概念带到现实中来的，让我们强烈地感受到这个地区的建筑的耀眼未来。

南街142号
桑迪·伦德尔建筑事务所

The brief for this project was to create a sustainable and contemporary family home that made the most of a dramatic but constrained brownfield site. Located on the banks of the River Ouse in Lewes, the site historically functioned as a wharf for the old quarry/cement works behind.

The house occupies a prominent position overlooking the river and nature reserve floodplain beyond, and is set against a backdrop of chalk cliffs to the south. The project aim was to design a unique home that respected its history and surroundings through form and material choice.

Sandy Rendel Architects

142 SOUTH STREET

The main body of the house is a simple two-storey pitched roof structure. It is reminiscent of the adjoining buildings but has a ridgeline that has been carved away to break down the building's scale and reflect the contours of the cliff face behind. In addition, the simple plan arrangement has been subtly distorted at each end to draw in key views and provide a covered terrace protected from the weather and buffered from the sound of the adjacent road.

A simple palette of self-finished materials was used to emphasise the buildings form. Each was carefully chosen to be robust, weather naturally, develop character, and reflect the tonal and material qualities of the site and surrounds.

On the ground floor an exposed frame has been constructed from board-marked concrete to reflect the tone and texture of the rugged concrete river wall below. Walls built from handmade glazed Sussex brickwork – a material traditional to the town – face South Street. This gives the home a softer texture and more intimate scale.

Above the muted tones of this masonry base, the first-floor walls and roofs are wrapped in a homogenous rain-screen skin that articulates the form of the upper volume. For this, the architects used a Corten steel expanded mesh, which weathers naturally into a striking red/ochre colour. This echoes the local red clay brickwork and tiles, whilst remaining distinct and alluding to the site's industrial heritage.

Traditional architectural elements such as gutters and eaves have been removed or concealed behind the mesh rain screen. This allows for crisp and clean detailing of the continuous surface and emphasises the home's primary form.

PROJECT SPECIFICATIONS

Location: Lewes, East Sussex, England
Area: 257 m² (2766 ft²)
Completed: 2015
Photography: Richard Chivers (pages 11, 12, 13 bottom, 15); Oliver Perrott (page 13 top)

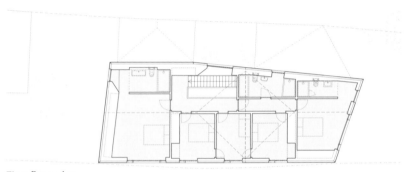

First-floor plan

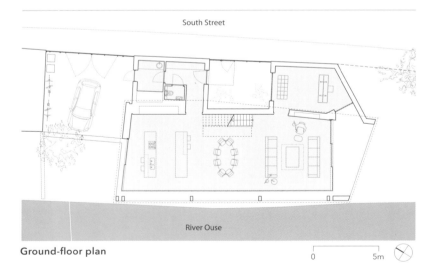

Ground-floor plan

安斯蒂·普拉姆住宅工作室
科平·多克雷建筑事务所

Ansty Plum is a home and studio in rural Wiltshire that has undergone an impressive retrofit and bold studio extension.

Comprising two imaginative buildings commissioned in the 1960s and 70s, the first is a one-bedroom house designed by David Levitt and the second, a studio and garage designed by the Smithsons.

The intensive repair, upgrading and re-organising of the two buildings brought about an 80 percent reduction in their energy use. The buildings are radically sited on a steep wooded hillside and overlook a collection of historic buildings.

Coppin Dockray Architects

ANSTY PLUM HOUSE + STUDIO

The brick and timber house has a simple open plan with a rectangular, singular-plane roof following the gradient of the land. The stone and concrete studio, hedged into the slope, peeps onto an ancient woodland track.

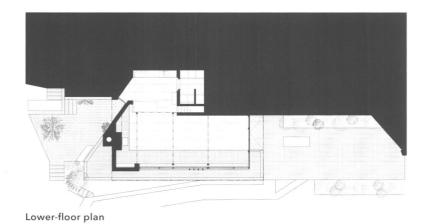

Lower-floor plan

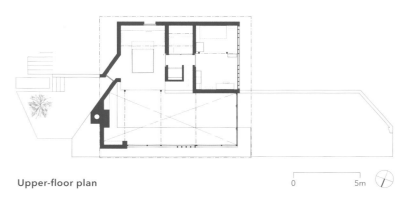

Upper-floor plan

0   5m

A number of changes had been made to the house over the decades, whilst the studio, after suffering structural failure, water ingress and decay, had been left derelict. Many accretive changes were removed to express the architectural qualities of the original house. The architects opened up the main space by removing a late addition bathroom and internal walls. They then created a new bedroom and study, and added central heating systems. As a result, the home once again displays clarity of intent and can be comfortably occupied throughout the year.

The architects also brought the studio back into use. The failed roof was replaced with an insulated zinc roof and cast-concrete copings. The structure was underpinned, tanked and insulated. Services and heating were also introduced. The studio now glows a warm pink colour from the meticulously detailed Douglas fir lining and joinery. The original structure was extended and hedged into the hill, creating a moody concrete and stone washroom that looks into a bank of prolific ferns. Access to the Smithsons' original working drawings allowed the architects to interpret many of the zinc, stone and timber details. This enabled the creation of a seamless extension that preserves the spirit of the building without compromising its functionality.

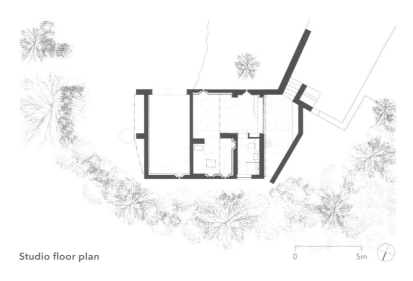

Studio floor plan

0　　5m

PROJECT SPECIFICATIONS

Location: Wiltshire, England
Area: 150 m² (1615 ft²)
Completed: 2015
Photography: Brotherton Lock

Chiles Evans + Care Architects

ARTEMIS BARN

阿耳特弥斯谷仓

奇利斯·埃文斯+凯尔建筑事务所

Located in the centre of Castleton in the Peak District National Park, Artemis Barn was in continuous agricultural use for over 250 years before falling derelict in the 1990s. The owner wanted to celebrate the barn's existing qualities whilst converting it into an exciting new home.

The main challenge was working with both the constraints and opportunities of the original building. Peak District National Park planning restrictions also gave very limited scope for new openings, particularly to the north wall, which is a long and blank façade.

The new plan challenges the idea of home: a living room without a view, a principal bedroom next to the main entrance and spaces for eating and relaxing defined within the openness of the existing volume. The materials reinforce the agricultural past of the building and a new reclaimed-stone roof punctured by a large light restores the original building's mass and drops light into the heart of the space.

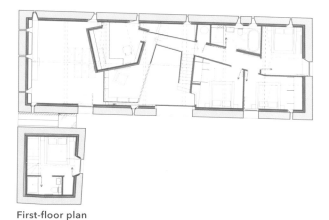

First-floor plan

Ground-floor plan   0   5m

The original rolling timber barn door has been minimally refurbished so when closed the barn retains its original appearance. It is a frameless glazed link to a small cottage that gives the structure's underlying modernity away. Internally, timber has been used in both rough and refined finishes. Concrete and steel are left in a raw state, and stone is exposed. The expressive, simple palette of materials reflects the skill of the local craftsmen who have worked on the project.

PROJECT SPECIFICATIONS

Location: Castleton, Peak District National Park, England
Area: 302 m² (3251 ft²)
Completed: 2016
Photography: Phillipa Evans (pages 23, 25, 26); Jen Langfield (pages 24, 27)

The house is at once intimate and vast with new spaces inserted into the major volume, each distinguished through form and material. The kitchen is flooded with light from an enormous glazed sliding door and the staircase is open and structural. Ever-changing daylight is animated throughout the structure, capturing the seasons through new and existing windows. At the first floor, the former hayloft level has three new bedrooms and a bathroom, whilst the space below houses a mixture of utility rooms as well as a single en suite and ground-floor bedroom.

27

奥斯丁住宅

亚当·尼布建筑事务所

This recently converted dwelling is situated on the site of the old St Swithun's School located along North Walls in Winchester. Vacated in 1929, the school had out grown the city centre location allowing the premises to be taken over by a library, followed by further development into residential dwellings. Austen House forms part of this old development.

The clients loved the location but wanted more natural light, a south-facing aspect and additional space. It was conceived that with an access way beside the property, which would only have to provide a right of passage for cars very

Adam Knibb Architects

AUSTEN HOUSE

occasionally, it was possible to build out and create a new room over the top of this space. The client undertook all necessary legal proceedings for permission of a flying freehold, whilst the architects created a very contemporary solution.

The home is located within the town centre conservation area, which meant its architectural addition had to stand out from the main building in both materiality and design.

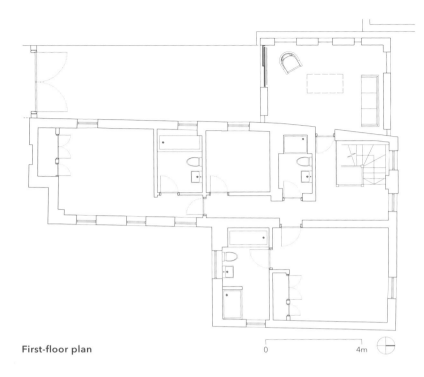

First-floor plan 0 4m

A floating timber box lightly touches the building via glass-slot windows and cantilevers out over the access way, providing dramatic internal space and making it a prominent feature down North Walls Road. Vertical timber cladding and large sections of glass have been specifically designed to give a greater feeling of verticality and sense of place. A sky lantern linked with actuator-controlled ventilation provides the space with additional aspects to the sky. The box stands delicately on two galvanised steel columns at the extent of the boundary.

The nature of the site meant principles of construction also had to be considered. The project was partially completed off-site to ensure minimal disruption on site.

PROJECT SPECIFICATIONS
Location: Winchester, Hampshire, England
Area: 21 m² (226 ft²)
Completed: 2015
Photography: Martin Gardner

Platform 5 Architects

BACKWATER

回水别墅

第五站台建筑事务所

This new detached home replaces an outdated bungalow on a promontory in a secluded lagoon in the Norfolk Broads. Designing a new building for the plot provided an opportunity to enhance the setting by establishing a stylish counterpoint to more traditionally designed neighbouring houses, whilst respecting the peaceful location. A key objective was to create simple, contemporary living spaces orientated to take advantage of surrounding views.

The house is arranged as three low-rise bays, each with pitched roofs echoing the working boat sheds typically found on the broads. Each bay has been positioned to address the waterfront, optimising natural light and providing internal living spaces with framed views across the water. The home's entrance is relatively low-key, with the main architectural drama reserved for the waterside elevation. Timber cladding offers a strong material contrast with neighbouring houses and is already beginning to weather back attractively, providing a sympathetic presence against the surrounding trees and water. Deep eaves emphasise the home's bold contemporary silhouette and provide sheltered external spaces, which are useable across the seasons. Externally, a layered timber landscape provides the transition from water level up to the elevated ground floor.

Inside, a simple broken-plan arrangement accommodates family life and allows for flexible living. The central bay contains a large kitchen and dining area, which flows into the adjacent double-height living and relaxation space. Bedrooms occupy the third bay and are connected by an impressive spiral staircase, which rises from the entrance hallway. Each space has a carefully considered interior with built-in furniture.

The peninsula location made construction a difficult task. In response, the project was designed with easily transportable, lightweight elements. Backwater is a highly sustainable family home, which brings a strong architectural presence to a sensitive waterside setting.

First-floor plan

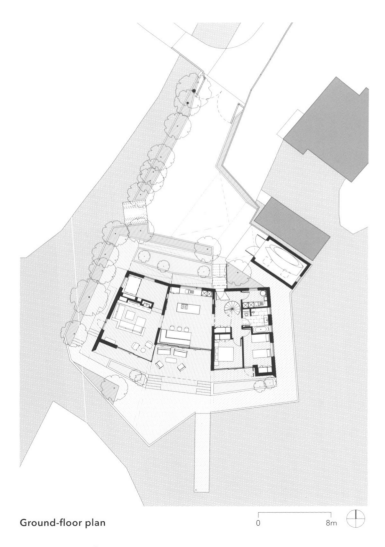

Ground-floor plan

0　　8m

PROJECT SPECIFICATIONS
Location: Norfolk, England
Area: House: 165 m² (1776 ft²); Boathouse: 20 m² (215 ft²)
Completed: 2016
Photography: Alan Williams

BALLYMAGARRY
公路住宅

2020建筑事务所

The concept behind this home was to rebuild and conserve an existing dwelling and old forge whilst creating a visual barrier for a modern, highly glazed, curved-roof living extension to sit on the block behind.

Two gables create strong basic shapes and replicate the agricultural buildings throughout the Irish countryside. The architects and planners wanted to retain as much of the original forge as possible and conserve its character, which would help to ground the design in the Irish vernacular. The curved roof allows the extension to sit lower than

2020 Architects

BALLYMAGARRY ROAD HOUSE

the original forge and addresses the need for two storeys. The two strong forms are divided by a flat-roofed section, which has been planted as a wildflower meadow. Due to the sloping nature of the site, the building increases in height along the road from west to east. This is an original feature used to enhance the drama of design internally.

The home is approached from the lower western gable. Upon entering, a 16-metre-long (52-foot-long) wall of natural unpainted stone pulls views into the depth of the house. Partially rebuilt using the original stone, this thick wall ties the modern structure to the site's history. The wall increases in scale when moving down a corridor created by the two competing building forms and wildflower roof above. The tension between these two spaces sets up much of the home's drama: the curving wall of the modern extension pushes against the heft of a massive stone element. To ensure its integration into the undulating fields beyond, the home sits on seven different levels.

The project incorporates a glulam-engineered superstructure consisting of five curved beams, which have been left exposed, making it one of the home's most prominent features. The beams break up the long cylindrical ceiling and provide a natural beauty and warmth to a largely blank canvas. Lowering environmental impact was also of great importance during the build and sourcing a locally made timber structure, as opposed to a steel structure or imported timber frame, was essential.

This home demonstrates the flexibility of design timber can provide. It is a project that uses structure not only as a means by which to support a roof and walls, but as an integrated component of the overall design aesthetic.

PROJECT SPECIFICATIONS

Location: Portrush, Northern Ireland
Area: 280 m² (3014 ft²)
Completed: 2015
Photography: Aiden Monaghan Photography

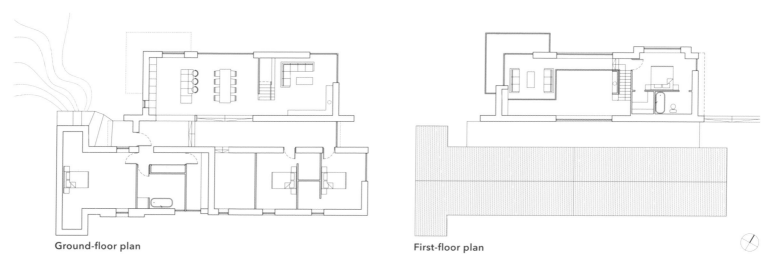

Ground-floor plan

First-floor plan

贝克特别墅
亚当·尼布建筑事务所

Tight planning constraints meant the architects had to take an inventive view towards attaining the space required for this whole-house refurbishment and extension. The original home's countryside location required the new design to make use of a planning loophole that only allowed a small dwelling to be extended by 25 percent if below 120 square metres (1291 square feet).

It was proposed that the existing 1950s farm worker cottage would be refurbished into a very contemporary house. The ground-floor extension gives the opportunity for a large vista view to the garden and woodlands beyond. At first-floor level a new master suite was added, which pokes out around the existing building and benefits from dual aspect views. The exterior concept was for a distinct datum between the ground and first floor – a solid monolithic masonry base at the ground and a first-floor timber box appearing to sit lightly on top. To avoid an overbearing presence from the road, the design takes a soft and modern approach at the home's front then completely transforms at the rear.

Adam Knibb Architects
BECKETT HOUSE

Internally, a radical floating staircase allows natural light to flow through the front hallway and into the rear open-plan space. Large slim-line sliding doors blur the transition between internal and external space.

PROJECT SPECIFICATIONS

Location: Crawley, West Sussex, England
Area: 198 m² (2131 ft²)
Completed: 2015
Photography: Martin Gardner

First-floor plan

Ground-floor plan 0 3m

布拉肯伯里住宅
尼尔·达什艾柯建筑事务所

This remodel and extension was designed for a young family with one child. The home forms part of a terrace of five Lillian Villas built in 1879.

The clients wanted a natural, earthy palette of materials and to retain as much of the existing building's fabric as possible, reusing what could not be preserved in the new reconfiguration They also wanted the house to feel connected to the outdoors by framing views to the garden and accommodating for as much natural light as possible. An open plan and sustainability were also important design factors.

Neil Dusheiko Architects

BRACKENBURY HOUSE

The design added a new basement and rear extension to the home, providing an expanded living and kitchen area; sky lit, sun-filled bathrooms; and a home cinema, playroom and guest bedroom. A contemporary rethinking of the basement typology ensured it would be light filled and multifunctional. The space acts as both a playroom for their young son and sophisticated cinema room.

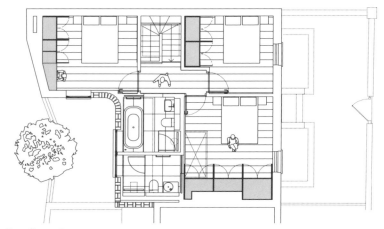

First-floor plan

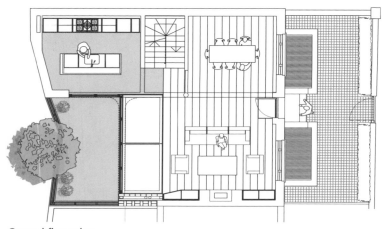

Ground-floor plan

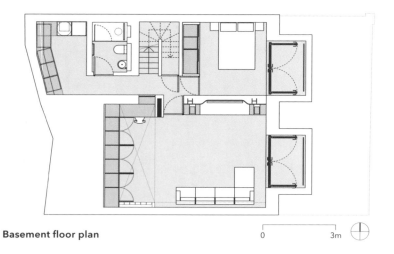

Basement floor plan

0 3m

A carefully selected, unified material palette creates calm and atmospheric spaces, stitching modern design into the historic context of the home's interior. Inherent relationships between re-used materials juxtaposed with modern industrial elements create a sense of aesthetic harmony throughout.

All levels of the home have been opened up to allow existing rooms to give way to an open-plan interior on the ground floor. In the new basement area, large skylights have been cut into the floors above to light the space. The kitchen and living rooms open out to a southwest-facing patio, creating a strong connection between house and garden.

PROJECT SPECIFICATIONS

Location: London, England
Area: 180 m² (1938 ft²)
Completed: 2015
Photography: Tim Crocker (pages 55, 56, 57 right, 58); Agnese Sanvito (pages 57 left, 59 bottom); Charles Hosea (page 59 top)

庭院住宅
达拉斯-皮尔斯-坤脱罗事务所

This award-winning two-bedroom house is arranged around a series of courtyards within the walls of an infill site in East London. The challenge with the L-shaped plot was to create a private house without compromising the neighbour's daylight or privacy.

Four distinct courtyards provide the main source of daylight for the house and open up sightlines through the property, creating functional outdoor spaces. The home's roof is at its lowest point at the rear of the neighbouring property. This sunken-roof design aspect minimises the home's impact within its surrounding context.

Dallas-Pierce-Quintero
COURTYARD HOUSE

The client wanted a home that would make best use of the site whilst allowing for flexible use of space. The result is an almost entirely open plan with separation and variety created with level changes and varying ceiling heights. A standalone study sits opposite the house. Within the space between, there is a walled kitchen garden, which provides the home with a green space, biodiversity, fruit, herbs, and flowers. The second bedroom is located on the ground floor as an extension of the living space and has been subdivided with a curtain. A rear patio provides a quiet space for entertaining and contemplation.

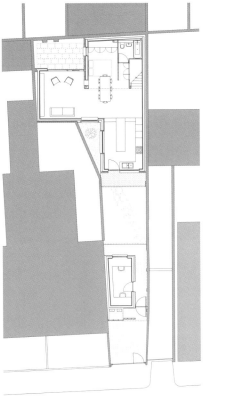

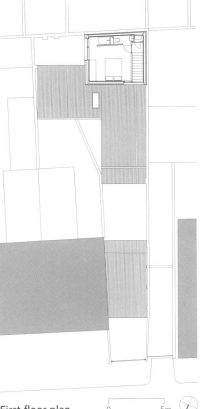

Ground-floor plan   First-floor plan   0   5m

Timber has been used for the home's structure and chosen for its cost, speed of construction and sustainability. Wall panels were fabricated off site and the roof structure built on site. Internally, the ceiling joists are exposed in the living areas to maximise height and add richness, changing direction as the plan kinks. A visible fire-rated board forms the underside of the ceiling.

63

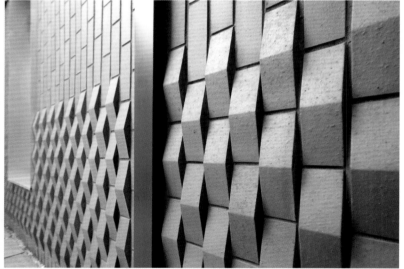

The building's materials were chosen for their affordability and resonance with the site. Black-profiled cement sheets have been used to clad the first-floor bedroom and roofs, echoing the corrugated steel on nearby outbuildings, whilst at ground floor, where the majority of the timber walls are protected by the existing brick boundary walls, the exposed walls are faced in blue brick or white render. Below waist height, sawtooth bricks have been used, softening the appearance of the brickwork.

PROJECT SPECIFICATIONS

Location: London, England
Area: 95 m² (1023 ft²)
Completed: 2014
Photography: David Butler (pages 61, 63 right); Rachael Smith (pages 62, 63 top left, 65); Tom Gildon (page 64)

Ashton Porter Architects

DRAG AND DROP HOUSE

拖拽别墅

阿什顿·波特建筑事务所

In response to a sloping plot, this project has been designed with a two-storey elevation to the rear and a stepped three-storey elevation at the front. The rear of the house has been embedded into the plot and a series of gabion walls provide a transition from the landscape into the main living spaces. These spaces are split into two main areas: kitchen/informal seating and formal living space. The stepping of the site continues through the central circulation space and exits through the front entrance, continuing out into the landscape.

A central circulation core, which connects all three storeys with an open and vertical void, divides the living spaces and sets up a series of rotating views throughout the home, some of which extend out into the plot and frame elements of the landscape.

By cutting into the sloping ground a lower-ground floor level was created, which acts as a retaining structure repeated as an artificial edge at the front elevation. This newly articulated retaining wall wraps into the landscape and acts as a mediator between the ground plane and the main floating volumes of the house above.

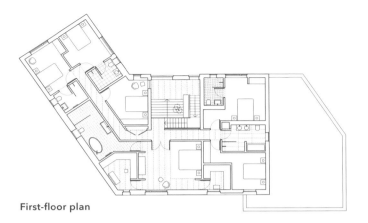

First-floor plan

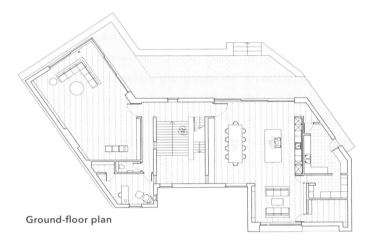

Ground-floor plan

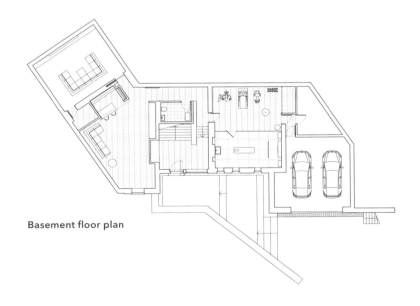

Basement floor plan

PROJECT SPECIFICATIONS
Location: North London, England
Area: 700 m² (7535 ft²)
Completed: 2016
Photography: Andy Stagg

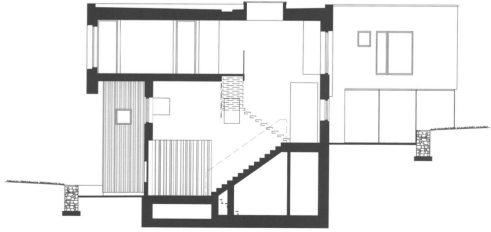

West section

逃至屋顶住宅
小工作室设计公司

This residential loft conversion of an Edwardian house includes a refuge for the parents of a growing family. The project brief was to design a place of peace and tranquility they could escape to after a long day's work.

The original four-bedroom family house had an ornate period front and a rear elevation with an apex-shaped closet extension. The aim of the new project was to re-create the apex geometry by making three careful incisions into the roof and inserting simple, modern and beautiful dormers to accentuate the Victorian geometry but in entirely different materials and using modern technology.

A Small Studio

ESCAPE TO THE ROOF

The dormers are only 250 millimetres (9.85 inches) apart, so the construction was challenging and required meticulous contractors. The dormers have full-height glazing finished in an apex shape and are framed in a black powder-coated aluminium trim. The dormer sides and roof are clad in vertical-pleated zinc.

Loft floor plan

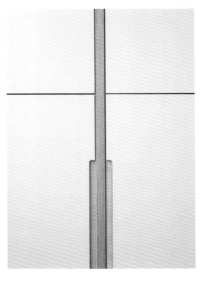

Internally, these windows frame views over southeast London and provide a busy backdrop to an otherwise minimal interior. The three spaces (bedroom, bathroom, wardrobe) were kept open-plan throughout the loft and have only been divided using a bespoke joinery system designed by the architect.

PROJECT SPECIFICATIONS

Location: South London, England
Area: 60 m² (646 ft²)
Completed: 2016
Photography: Jim Stephenson

Space and light were critical in the design. In order to maximise natural daylight throughout the house, a central void was cut through the entire structure. The new void has an apex ceiling construction and houses a new freestanding solid oak staircase, which lands gently on the first floor of the original house, respecting its period features.

Bureau de Change

FOLDS HOUSE

折叠住宅

外币兑换所建筑事务所

The focal point of this project is a pleated roof at the back of house. It appears to be formed from a flat surface and forced to crinkle up into a faceted structure as it's pushed up against the exterior wall. From the garden, the pleats are purposefully sunk from view, creating the impression of a simple flat roof, which allows the character of the original building to stand out. The design intention was to evoke a graphic feel within a modern extension, creating a feeling of motion that would emphasise the meeting of old and new.

The poise of the roof offers natural points of placement for generous skylights. This creates light within the living area and satisfies the owners' desire to see the nearby woodland whilst relaxing in the space.

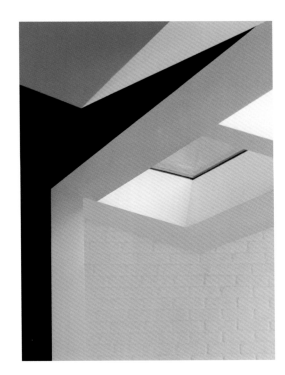

Forming a side and rear extension, the roof expands the existing kitchen and creates a new dining and work area. The end of the pleated ceiling, which is capped by a midnight blue surface that emphasises its ample peaks and troughs, marks the boundary of the kitchen assertively. Darkened surfaces continue through the kitchen into a long passage that extends down the left side of the property, conveniently and discreetly creating a storage and utilities zone.

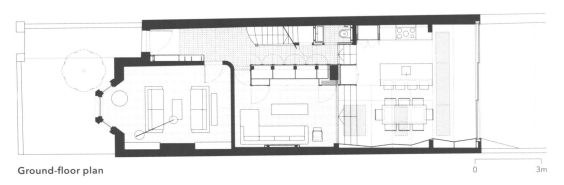

Ground-floor plan

PROJECT SPECIFICATIONS

Location: North London, England
Area: 40 m² (431 ft²)
Completed: 2015
Photography: Bureau de Change

The home's scheme is characterised by a rich palette of colours, materials and textures, which create a different experience in each space and mark transitions between living spaces. Large terrazzo slabs, encaustic tiles and tonal parquet complement the complexion of the original building, whilst enhancing the graphic impact of the angular extension.

画廊住宅

尼尔·达什艾柯建筑事务所

This project design added a light-filled side extension at ground-floor level and a new loft bedroom and bathroom in previously unused roof space to a home in need of modernisation.

By opening up the front and rear reception rooms an open-plan living space was created, providing ample display space for the owner's extensive art collection. A fully glazed skylight to the side extension allows for increased light levels and improved connections between spaces linking the front of the house to the rear garden.

Neil Dusheiko Architects

GALLERY HOUSE

A set of new stairs run full width across the structure, marking the connection between the new and old rooms. Niches have been carved into the depth of the walls to allow for storage and spaces for sculptures to be displayed.

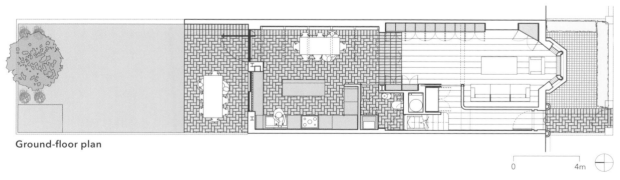

Ground-floor plan

0 4m

A calm material palette of reclaimed bricks and oak flooring gives a sense of warmth and texture to the home. The rich detailing gives a tactile scale to the new domestic spaces. Built-in kitchen seating provides and informal gathering space that looks out into the garden. A large glass-pivoting door enables generous access to the same outdoor space.

A framework of oak shelving and timber rafters gives structure to the new ground-floor space, providing further display space for art, ceramics and glassware pieces under a bright sun-filled, fully glazed skylight.

Section

0 4m

The new loft level is a timber framed zinc-clad structure, which adds an additional bedroom and bathroom to the property. It also allows for panoramic views towards Clissold Park and the distant church steeple.

PROJECT SPECIFICATIONS
Location: London, England
Area: 140 m² (1507 ft²)
Completed: 2016
Photography: Agnese Sanvito (page 85); Tim Crocker (pages 86, 87, 88, 89, 90, 91)

Ström Architects
HARTROW

哈特罗别墅
电流建筑事务所

Hartrow is a large-scale refurbishment and extension project of a 1960s house in Winchester. The original house was unusual in its residential street setting, presenting an entirely different aesthetic, orientation and layout to its neighbours. Part of the challenge was to ensure that the new design would celebrate its mid-century styling whilst bringing its function and performance up to 21st-century standards.

The house was split over several levels to suit the sloping site, with a drive and entrance hall to the street. Externally, the driveway ran steeply alongside the house to the carport at the rear, several levels below. The living rooms and kitchen were located on the top floor. This arrangement of space no longer worked for modern life and the multiple levels felt disconnected from one another and the site.

The new internal layout was arranged to fulfill the needs of a busy modern family. The unused carport was enclosed with sliding glazing and turned into a family room for cooking, playing, eating, and relaxing. This created a relationship with the exterior, allowing the space to open up to the garden for summertime entertaining and supervised play space for the children. A calmer, more sophisticated living room is located on the top floor, taking advantage of views and providing a break from the informal family space on the bottom floor. Bedrooms, a television room and a study are housed on the intermediate floors.

In order to create a sense of connectedness between the five split floors a void was punched through the floor plates. This provides views, light and acts as a reference point on each floor.

Building fabric was updated to make the house more energy efficient and robust. The external felted finish of the original roof was removed and a crisp black-zinc roof added as a durable and contemporary replacement. The internal balsa wood, a prominent feature of the home's original design, was retained and carefully refurbished.

Despite the relatively superficial nature of the works carried out, this new design succeeded in bringing an ageing house into the 21st century.

Upper floor plan

Intermediate floor plan

Lower floor plan

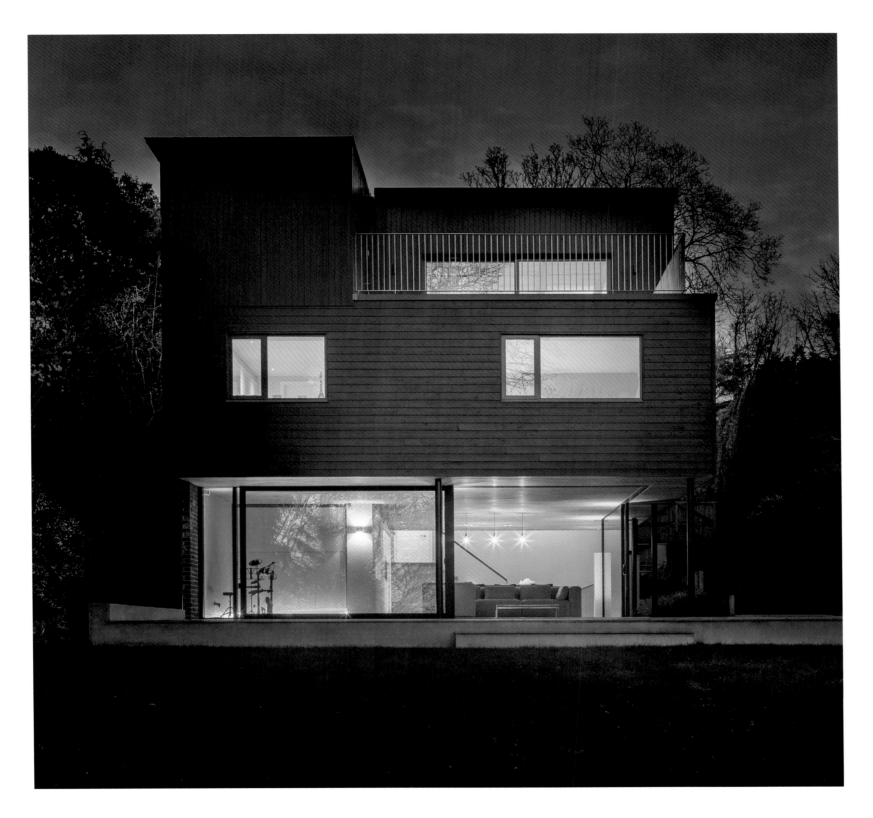

PROJECT SPECIFICATIONS

Location: Winchester, Hampshire, England
Area: 460 m² (4951 ft²)
Completed: 2014
Photography: Martin Gardner (pages 93, 94, 95, 96); Mike French (page 97)

黑兹尔巷马厩住宅
都柏林设计工作室

This scheme places three two-storey contemporary dwellings onto a vacant plot.

Hidden from the street, a 5.5-metre-wide (18-foot-wide) laneway located between two semi-detached dwellings grants access to the site. The site is 70 metres (230 feet) long and just 10 metres (33 feet) wide, and shares garden boundaries with seven dwellings to the north and a laneway to the south. Issues of privacy, the site's constraints, and overshadowing dictated the project's geometry, orientation and size.

Dublin Design Studio

HAZEL LANE MEWS

The core project concept was to maximise the available site area of each house without impacting on the amenity of any of the adjoining dwellings. To achieve this, the footprint of each site, including the garden and house, was conceived as a single entity.

The garden is an integral part of the overall ground-floor plan because it mirrors and interlocks with the house plan. By introverting the external garden area into the design of the house, the site area is maximised without any impact on the adjoining dwellings. This concept allows the ground floor to be as open as possible, taking advantage of the site's southwest orientation. The laneway façade is designed as a defensive or boundary wall, pierced only by small openings with timber louvre screens. On the garden side, in sharp contrast, full-height glazing allows maximum penetration of light deep into the plan from the south and west façades.

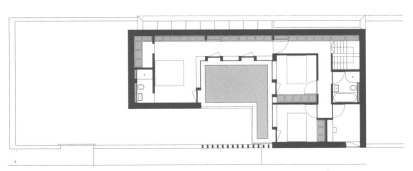
First-floor plan

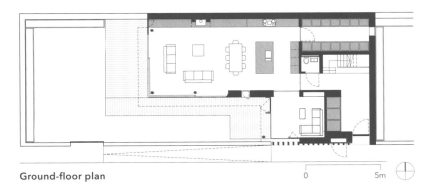
Ground-floor plan

0　　5m

The first-floor level has been treated similarly to the ground floor, with all windows focused around a planted sedum roof. This south-facing roof garden and full-height glazing allows light to penetrate deep into the bedrooms without the possibility of direct or indirect overlooking from adjoining properties. The Iroko louvres at the ground floor extend up to the first-floor level to form a privacy screen.

The openness and transparency of the south and west façades at the ground-floor level juxtapose sharply with the closed treatment of the east and north façades, which have been left as solid brick planes to ensure no overlooking of adjoining gardens. The first-floor brick-clad box accentuates the solidity of these inward looking sculptural volumes. The uncompromising outward façade projects a determination to ensure the privacy of those inside and of those in surrounding houses.

PROJECT SPECIFICATIONS

Location: Clontarf, Dublin, Ireland
Area: 180 m² (1938 ft²)
Completed: 2015
Photography: Alice Clancy

Jestico + Whiles
HOUSE 19

19号住宅

Jestico + Whiles 建筑事务所

This project is a carbon-neutral, energy efficient home that fuses traditional forms and local materials in an elegant, modern way to make full use of the natural aspect and orientation of its site.

An entrance gallery leads to the main body of the house, with bedrooms on the upper level and large picture windows framing considered views. The house modulates daylight internally to dramatic effect, resulting in a tranquil atmosphere that changes throughout the seasons. Kitchen, dining and living spaces combine to create a family hub with visual and physical connections to the south-facing garden. A single-storey studio has the flexibility of being used as a garden room, office or ground floor bedroom.

Practical and deliverable sustainability features form the bedrock of the design. The home features a ground source heat pump, earth tube ventilation, photovoltaic panels and rainwater harvesting. A double-height space at the heart of the plan works at an architectural level uniting the two floors, but is also a passive stack, allowing high-level vents that open during summer to create natural cooling.

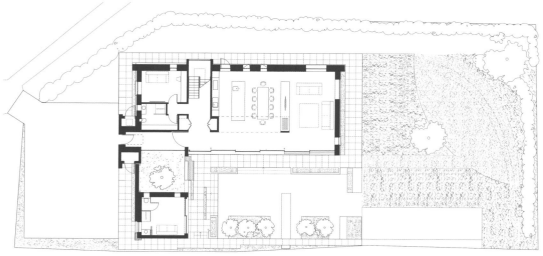

Ground-floor plan

First-floor plan 0 5m

Materials are locally sourced and firmly rooted in the history of the area. Black-stained cedar cladding echoes agricultural barn buildings; the local church inspires the snapped-and-knapped luminescent flint; and Corten steel accents have been chosen to reflect rusty deposits found on the flint.

Expressing the two storeys on the northeast elevation and reducing the height on the southern elevation to a single storey by means of an eccentrically pitched roof, means that the impact on the adjoining property is minimised. This, together with the materiality of the house means it adopts a commanding yet respectful presence in the context of its neighbourhood.

PROJECT SPECIFICATIONS
Location: Old Amersham, Buckinghamshire, England
Area: 246 m² (2648 ft²)
Completed: 2016
Photography: Grant Smith

艺术爱好者住宅

shedkm 建筑事务所

This house was conceived as a simple two-storey villa placed in the centre of a flat rectangular site surrounded by mature trees. The ground-floor spaces are defined by a series of interconnecting white-painted brick walls, with full-height glazing between. The plan of the house and arrangement of these walls divides the garden into four quadrants, each with a different theme.

Open-plan kitchen and dining spaces are at ground level and arranged around a top-lit, double-height stairwell, which also acts as a gallery space. The gallery bleeds into adjoining spaces with paintings carefully positioned in each. These spaces have strong connections to formal and informal garden areas. The upper level, a timber-clad volume with screened windows and carved out balconies, contains the family's private quarters and includes bathrooms lit only from above.

shedkm

HOUSE FOR AN ART LOVER

Externally, the home is clearly articulated with a simple palette of materials. Whitewashed brick, Iroko cladding and timber-framed glazing bring scale and a more domestic feel to the large volumes. Careful arrangement of formal and informal planting bed the house into its setting.

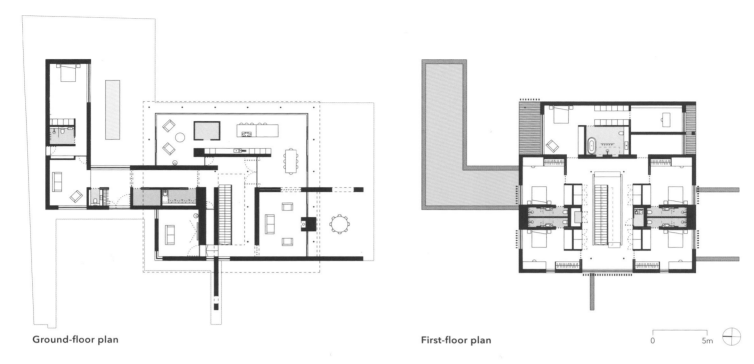

Ground-floor plan　　　　　　　　　　　　　　　First-floor plan　　　0　　5m

PROJECT SPECIFICATIONS

Location: Merseyside, England
Area: 650 m² (6997 ft²)
Completed: 2014
Photography: Jack Hobhouse

痕迹之屋
鹤田建筑事务所

After the demolition of an original extension, this home required an intervention that would honour the original building whilst simultaneously allowing the new project to have its own identity.

The original extension had no distinct historical or architectural value and was structurally unsound. The architects chose to incorporate the home's sloping roof – a typical terrace house feature – into the new face of the rear garden, preserving its charm and a sense of memory. The new envelope's structures have been internally exposed wherever practically possible, with the intent of registering future stories of the house.

Tsuruta Architects

HOUSE OF TRACE

The hand marks of bare plaster finish left exposed in the bedrooms are now recorded on the internal faces of the building's fabric. The slow patination of bespoke copper and brass fittings reflect a passage of time as they change from their original colour. In response to the home's increased depth, a two-storey lightwell was added between the new and original part of the building, bringing daylight to interior rooms. The lightwell serves as a focal point in the home, connecting the family between floors (open kitchen, dining level and bedrooms).

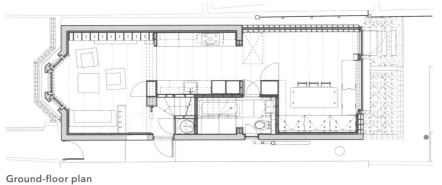

Ground-floor plan

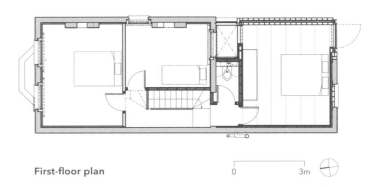

First-floor plan 0 3m

On the upper floor, the master bedroom and children's room each has an internal window facing the lightwell. This gives the parents and children a sense of one another's presence whilst also allowing for privacy when desired.

The entire building and its new components were three dimensionally modelled. Some components, including the main stair and furniture were further broken down and processed with CNC (Computer Numerical Control). They were then brought to the site as flat-pack packages, which significantly cut project cost and time. A record of the CNC flat-pack process was deliberately left visible on the faces of each piece, retaining memories of assembly within the home long after project completion. Materials used include: facing bricks, exposed steel, white cement concrete, aluminium, timber, composite frame windows, pine-board panels, birch plywood, melamine plywood, coloured MDF (solid), glazed subway tiles, and copper pipes with brass garden taps. The project reveals memories of place and registers future stories, allowing the past and present to coexist in a brand new home.

PROJECT SPECIFICATIONS

Location: London, England
Area: 110 m² (1184 ft²)
Completed: 2015
Photography: Tim Crocker

赫德尔住宅
亚当·尼布建筑事务所

Adam Knibb Architects were approached to put together a contemporary extension for a Grade II–listed house in Alresford. Hurdle House has a long history within the village, having been part of the original sheep fairs back in 1792. The clients wanted a contemporary addition that would maintain respect for the original barn.

Adam Knibb Architects

HURDLE HOUSE

The scheme was approached with an aim to set works into the surrounding nature, provide natural light, harness the fantastic views, and provide a social heart to the house. Working with the Winchester conservation department, it was agreed that the rear bay window could be removed to provide a linking element to the extension. A frameless glass link was envisaged, touching the existing building lightly and connecting the old to the new. The extension comprises a large open-plan kitchen, dining area, casual seating with utility and water closet, and attached study.

A major aim of the project was to increase excitement upon entering the property. The new main entrance directly introduces this proposal, showing off beautiful views of the garden. Vertical timber cladding provides a contemporary contrast to the existing building and mimics the surrounding trees. With large glass apertures, the property's natural surrounds blur boundaries of the extension, making the home a prime example of contemporary architecture at harmony with the natural landscape.

PROJECT SPECIFICATIONS
Location: Alresford, Hampshire, England
Area: 110 m² (1184 ft²)
Completed: 2016
Photography: James Morris

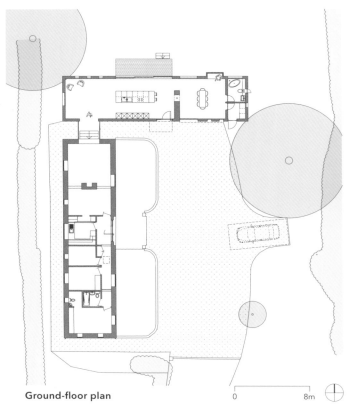

Ground-floor plan 0   8m

无限住宅
迷幻设计公司

Theatre, drama and illusion are central to the architecture of Infinity House.

It is an exciting sequence of contrasting spaces, which interconnect in a variety of ways to create a unique spatial experience. This design led, rejuvenation project substantially extends an 18th-century Georgian townhouse after removing a poor quality industrial workshop that covered the original garden.

Spaced Out Ltd
INFINITY HOUSE

The sanctuary and classic section of the townhouse has been seamlessly linked vertically and horizontally to new spaces. Soft reflections and floating details deliver transitions between the old, new and nature. The structured planting of sky gardens, visible from all main spaces, invites nature into the home.

The home's glazing is impressive, with Hepco Motion applying its automated sliding technology to create the UK's widest vertical motorised sash window. Glazing has also been used in other areas of the house, enabling privacy or interconnectedness at the touch of a button. The home boasts a variety of dramatic and multi-directional views, connecting features such as the swimming pool to the sky.

Smart controls and recyclable energy have been implemented throughout the home. This enables efficient and easy-to-run spaces whilst seamlessly combining traditional architectural craft-based techniques with digital technology. Privacy, as an example, is managed with carefully selected planting and traditional shutters alongside interactive glass interlayers.

There are numerous special moments, great spaces and details that create the undeniable theatre of this unique London townhouse.

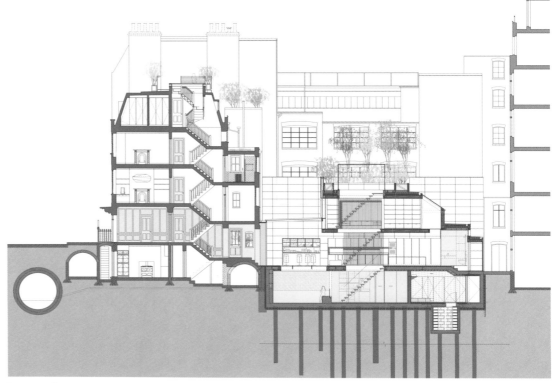

Long section

132

PROJECT SPECIFICATIONS

Location: London, England
Area: 500 m² (5382 ft²)
Completed: 2016
Photography: www.joshphoto.com

诺尔住宅

奇利斯·埃文斯+凯尔建筑事务所

This project radically transforms a 1950s house into a unique and highly efficient home suitable for a young family. The client's brief sought to create a low energy home that was architectural yet restrained. New insulation and bricks, which envelope the building, improve the home's thermal performance and give it a fresh identity.

Careful and considered brick detailing reflects the client's fastidious attention to detail and helps to control façade proportions.

Chiles Evans + Care Architects

KNOWLE HOUSE

From the street, the house is deliberately restrained: a simple two-storey mass of brickwork retaining many of the window openings of the original house, capped by a slate-clad roof. A low-sweeping, timber canopy projects from the front of the house, suggesting that the external form belies the internal experience.

The house is entered from street level via a low, timber-lined entrance. This gives way to a triple-height stairwell, which links all of levels of the house. At the first-floor level, ceilings follow the new roofline, creating a dynamic conclusion to the stairwell. The panoramic window of the lounge space on the upper-ground level reveals the dramatic topography hidden from the street. The deep timber reveal creates a space to sit and look over the garden below as well as towards the white peak beyond.

The lower-ground floor forms an enfilade of rooms designed to allow for different activities to take place in adjacent spaces and to bring the family together.

Deeply recessed, glazed elements frame views across the wide terrace and over the trees at the foot of the garden. A herb garden and raised lawn complement the terraced area and further blur visual edges with the wider landscape. Generous brick-lined steps lead to the informal garden below.

PROJECT SPECIFICATIONS
Location: Yorkshire, England
Area: 1890 m² (20,344 ft²)
Completed: 2016
Photography: Jeremy Phillips

Lower floor plan

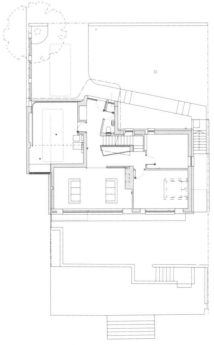

Ground-floor plan

First-floor plan

Hudson Architects

LE PETIT FORT

小堡别墅

哈德森建筑事务所

This home sits on the shoreline of St Ouen's Bay and occupies the site of an earlier, now demolished, farmstead. The original building was constructed in the early 20th century and enclosed within thick granite walls, which have been retained and restored. The house was conceived to complete the 'fort' concept by building the missing fourth wall of the enclosure and creating a central element to represent a 'keep'.

The three-storey entrance block, like the perimeter walls, was constructed from Jersey granite and reclaimed from the earlier building. A pair of two-storey wings radiates from this block, dividing the walled enclosure into an entrance forecourt and sheltered garden, and enclosing a landscaped pool terrace. The two wings use a contrasting palette of contemporary materials: polished micro-cement render down to the ground floor sits below an oversailing and highly textured first floor of Corten and Iroko panelling. The first floor shelters the entrance and ground floor terraces, whilst the deep eaves shade a first-floor balcony and enhance the building's distinctive silhouette.

The composition offers a lively series of contrasting profiles from different viewpoints. From the entrance the fort-like character dominates, picking up references to nearby Napoleonic Martello towers and Second World War fortifications. Small windows set into the Corten panelling give the house a robust and intriguing presence. From the beach, the house appears as a low-set watchtower, with the keep and first floor peering over the top of the granite walls. From the pool terrace, the building assumes a more welcoming and transparent character.

PROJECT SPECIFICATIONS

Location: St Ouen's Bay, Jersey, Channel Islands
Area: 472 m² (5081 ft²)
Completed: 2015
Photography: Joakim Boren (pages 141, 142-3); Edmund Summer (pages 144, 145)

Inside, the main living area occupies the first floor of the sea-facing wing, with impressive views across the bay through full-height windows, which lead onto a balcony and terrace. The open-plan space is broken up by level changes: low dividing walls and a feature fireplace facing two sitting areas. The master-bedroom suite in the opposite wing is linked by a large landing, which provides additional relaxation space. The ground floor contains family bedrooms, a media room, sauna, laundry, and plant room. The main floors are linked by a stunning spiral staircase, which is lit from above by a conical lantern rising through the 'keep'. A smaller yet similarly crafted concrete spiral staircase leads to a study and guest-bedroom suite at the top of the house.

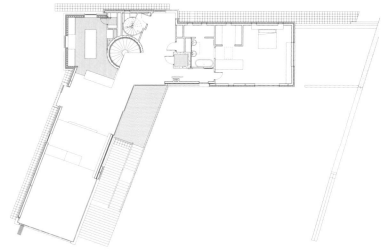

First-floor plan

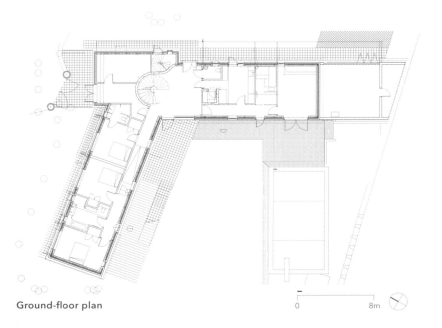

Ground-floor plan

狮门花园
3S建筑设计公司

Located in the heart of Kew conservation, this typical detached property was built to mimic the traditional Edwardian houses next door. The home lacked the detail and craftsmanship expected of homes within historical vicinities. Its dormers were set too high, denying garden views; the main staircase opposite the entrance hall appeared dark; and several level changes within the home gave a dated impression.

3S Architects and Designers
LION GATE GARDENS

The architects worked with the owners to fundamentally improve the home's interior flow. A large rear extension was added to the home. This allowed for a spacious living area with a kitchen at the upper floor leading down into a dining area.

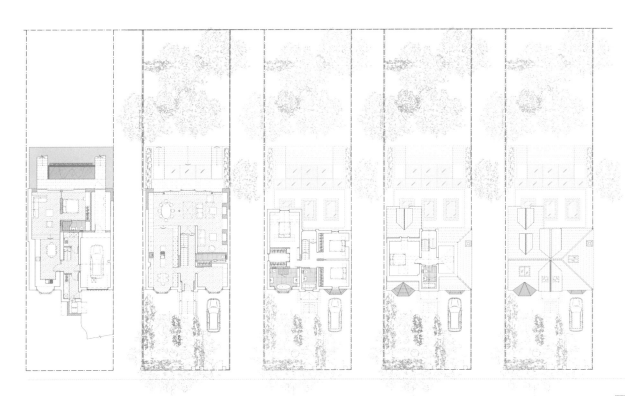

Basement floor plan   Ground-floor plan   First-floor plan   Second-floor plan   Roof floor plan   0   8m

The low ceilings of the existing basement were replaced with higher ceilings and a separate guest bedroom with moving walls. This space can also be used as a study/lounge, second kitchen or storage utility.

The upper bedrooms have been practically designed, with ample storage and spacious bathrooms. These spaces connect to a new stairway, which winds along an all-glass skin down to the basement. A clever mix of styles – New York loft meets Philip Stark – gives the home a feeling of city life.

PROJECT SPECIFICATIONS
Location: London, England
Area: 77 m² (829 ft²)
Completed: 2015
Photography: Alistair Nicholls

MAP住宅
SAM建筑事务所

Although the initial site for this home was particularly small – 7 metres by 7 metres (23 feet by 23 feet) – the architects sought to devise a beautiful design that would work within the unique London location. The aim was to create a home that would fit coherently within area's historic urban fabric, whilst being unashamedly modern and adhering to the owners' lifestyles.

SAM Architects

MAP HOUSE

The external fabric is a traditional cavity wall construction made of 150-year-old reclaimed dark bricks and special order black mortar. The bricks are laid with wide flush joints to create a continuous surface emphasising the joint instead of the brick. Minimal windows, gunmetal grey in colour, blend in with the tones of the walls and reinforce the perception of a simple and unitary aboveground mass.

First-floor plan

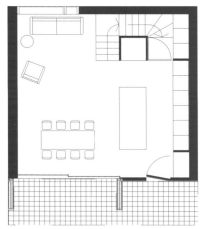
Ground-floor plan

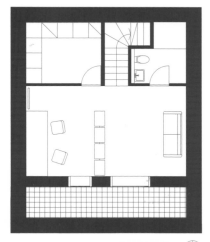
Basement floor plan 0 2m

On the mews side, the ground floor has been clad in charred-larch slats, which counterpoint the solidity of the brick volume with a lightly articulated vertical texture. A large part of this façade is an automated bi-folding gate, which opens to reveal a 3-metre by 4-metre (10-foot by 13-foot) glass wall, allowing the living space to be at one with the outdoors on warm summer days. From the inside, the open gates frame a view to the gardens and provide a sense of privacy from passers-by.

PROJECT SPECIFICATIONS

Location: London, England
Area: 49 m² (527 ft²)
Completed: 2015
Photography: Edmund Sumner

An open-plan kitchen and dining area, with an open stair leading to the first floor welcome entry into the home. This stair is topped with a large roof light, which brightens the ground floor. On the first floor there are two bedrooms and a generous bathroom. From the kitchen area, the lower-ground level can be accessed via the stair through a hidden door. The basement is provided light by a 1.5-metre-wide (5-foot-wide) light well with a galvanised grill, which sits at the front of the house. This area also includes a guest room, shower room, utility room, and study.

Materials used include oak, concrete, white plaster, charred timber, and bricks. Charred timber and bricks were the only materials used externally. The bespoke kitchen was designed as a workshop with very simple and hardwearing materials, such as plywood and linoleum.

橡树巷别墅

凯松·卡斯尔建筑事务所

This bespoke home replaces a small existing house, which had become neglected and incompatible with contemporary family life. The design approach was to derive a modern aesthetic from local vernacular architecture whilst maintaining a fundamental commitment to issues of sustainability.

The external finishes have been horizontally arranged with black timber at the first floor stacked onto a plinth of knapped flint. These materials are present in vernacular buildings nearby. The first floor is articulated as a series of boxes, which shift back and forth, providing the ground floor with roof lights and solar shading.

Cassion Castle Architects

OAK LANE HOUSE

Approached via a northern courtyard, the building initially appears as somewhat impenetrable and monolithic, but upon entering the interior reveals itself to be surprisingly light and open with a strong sense of transparency. Interior spaces were designed as a blank canvas so that the client – a keen interior designer – could make her mark over time. Polished concrete, oak parquet, natural stone, and a simple decorative palette give the house a timeless and elegant identity.

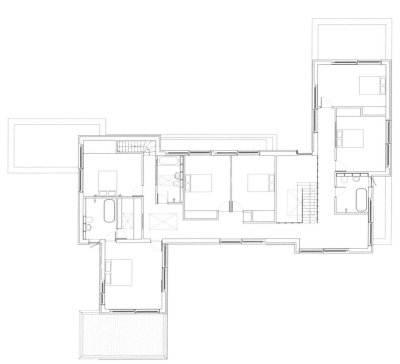

First-floor plan

Ground-floor plan

0   5m

161

Steel frame construction enabled a fast build and permitted the building's many cantilevers. A storey-height truss incorporated into the wall at first-floor level enabled the living room's wide, clear spans.

Inspired by local farm buildings, the ground floor has been separated into wings, which define external spaces around the home. These intermediary spaces create a relationship between the building and landscape: a welcoming north-facing entrance courtyard, a contemplative orchard, a large south-facing open garden, and a smaller west-facing children's garden. A double-height entrance hall leads to a generous central wing, which is broken into clearly defined zones for cooking, dining and relaxing. Large sliding doors disappear into cavity walls to provide a seamless transition from the interior through to the terrace and into the garden. The ground floor also contains a library, study, gym, playroom, utility, and boot room. Six bedrooms and bathrooms are accommodated within the timber blocks upstairs.

The home was designed to be highly sustainable. The first floor overhangs the ground floor and is set back on the north elevation. This creates a north-facing roof light and prevents overheating in summer. Rubble from the existing building was recycled and used for the new house. Other energy saving measures include photovoltaic panels, low-energy fixtures, a woodchip burning stove, and a biomass boiler powered by a sewage processing bio-digester.

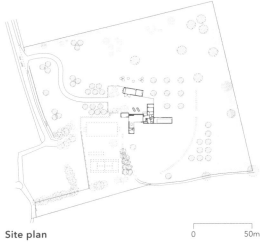

Site plan 0 50m

PROJECT SPECIFICATIONS
Location: Suffolk, England
Area: 450 m² (4844 ft²)
Completed: 2015
Photography: Kilian O'Sullivan

165

奥克兰斯住宅
布莱德斯通建筑事务所

Broadstone Architects

OAKLANDS

A primary characteristic of these particular mid-20th century houses is the large rear façade chimney. Given the height and presence of this home's chimney stack and its balance with the neighbouring equivalent, the architects decided to make it a focal point in the roof design, off which all extending roof pitches would generate.

In order to utilise the western aspect along the boundary of the sizeable rear garden, an extension stretches into the garden. This maintains light, ventilation and views to and from the original rear façade via a courtyard. Conscious of the neighbour's daylight aspect and long extending building, the architects formed the roof shape with low pitches, which spring from the main existing chimney stack. These low-rising angles are brought further into the extension to direct light; similar to how the roof rises to capture clerestory light.

The single-storey extension wraps around a new courtyard, allowing south and west light into the centre of the plan, whilst catching glimpses of the long landscaped garden. An external fireplace and stone gully for a rainwater spout add aspect to the courtyard, which is accessed from the kitchen and dining areas. A living room terminates the extension and opens into the garden, availing evening light with south-facing clerestory windows above the boundary wall. The bevelled canopy and edge wall enclose and hide utility access to the garden.

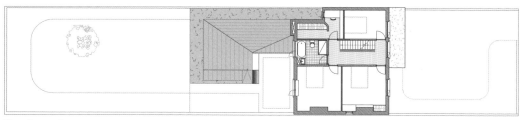

First-floor plan

Ground-floor plan

The kitchen is central and acts as a hub from which all areas can be be accessed. A full refurbishment of the main house provides a new staircase with change of direction. This allows daylight into the hall and provides access to the attic. Materials chosen include post-finished concrete and limewashed birch plywood, which complement the original industrial-style steel windows.

PROJECT SPECIFICATIONS
Location: Dublin, Ireland
Area: 204 m² (2196 ft²)
Completed: 2016
Photography: Aisling McCoy

旧水塔住宅

格雷斯福德建筑事务所

The Old Water Tower is a passively certified four-bedroom family home on the edge of Chieveley in West Berkshire.

The project was conceived as a modern interpretation of the timber-framed barns within the area, those that when seen at a distance appear as simple, traditional and agricultural buildings. On closer inspection, The Old Water Tower reveals itself as a crisply designed, carefully made contemporary interpretation of the historic buildings it draws inspiration from.

Gresford Architects

THE OLD WATER TOWER

The architects aimed to create a home that would eschew austere interpretations of eco-friendly living, whilst remaining on the cutting edge of sustainable design. The result is a home that is extremely easy to live yet costs virtually nothing to run. Space on the roof for photovoltaic and solar thermal panels provides the option of total self-reliance.

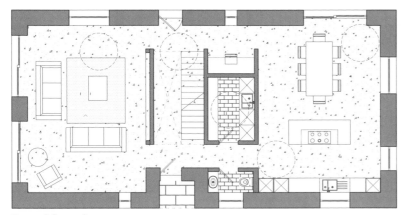

Ground-floor plan

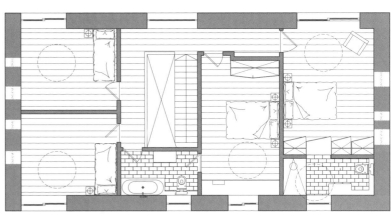

First-floor plan

PROJECT SPECIFICATIONS

Location: Chieveley, West Berkshire, England
Area: 173 m² (1862 ft²)
Completed: 2015
Photography: Quintin Lake

Another critical aspect involved in passive house construction is the internal level of comfort a home provides year round. This was achieved by installing a mechanical ventilation and heat recovery unit, which provides constant background ventilation by removing stale air and drawing in fresh air. This creates a fresh and ventilated feeling within the home, whilst maintaining warmth.

The house has openable triple-glazed windows and large triple-glazed sliding doors. To reduce the risk of overheating, external blinds on the east, south, and west elevation automatically deploy when the home's internal temperature reaches 21°C (69.8°F).

一上二下住宅

麦卡洛·姆尔文建筑事务所

One Up Two Down is an urban courtyard house in central Dublin built on a very tight budget. The site was carved out of an existing plot along the now in-filled Royal Canal.

The home's scheme fills the rectangular site and maximises open space. Loft-like living space on the upper floor allows for privacy, light and views. Below are bedrooms and a studio, which sit beyond the central courtyard. Nature is at the heart of the house. A walled and stepped garden planted with vegetables, grasses and wild flowers is located at the front of the home. A planted central courtyard, roof terrace and upper-roof sky garden provide additional green spaces for the owners to enjoy.

McCullough Mulvin Architects

ONE UP TWO DOWN

At the home's entrance, a light-filled staircase leads to the main living space and stretches between the island kitchen and seating area, directing views out toward the studio terrace.

Bricks from the original ruined house were salvaged and re-used to anchor the front façade. A screen of wooden Iroko ribs over a glass and timber elevation, giving the front of the home a striking identity. The ribs continue along the upper level of the central courtyard and have been angled to bring in sunlight and screen views. Internally, finishes are austere and simple – polished concrete floors, white-painted exposed rafters, and Iroko windows and doors.

PROJECT SPECIFICATIONS

Location: Dublin, Ireland
Area: 170 m² (1830 ft²)
Completed: 2016
Photography: Courtesy of the architect

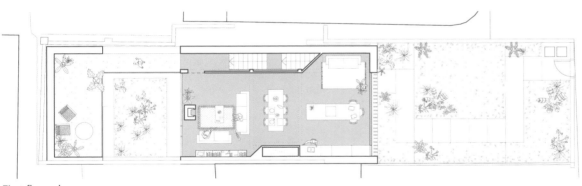
First-floor plan

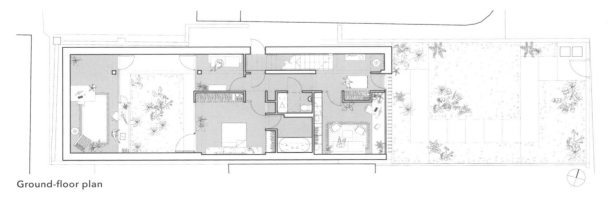
Ground-floor plan

183

页面巷住宅
柯克伍德·麦卡锡

Kirkwood McCarthy
PAGES LANE

A new ground-floor wing and loft dormer transform the layout of this Victorian semi-detached home to suit contemporary family living. Resisting the approach to demolish and achieve one large living space, the layout integrates the original external walls into a new internal layout, forging a new story-telling vernacular to the residence.

The footprint and form of extension elements is decisive and controlled. The new ground-floor wing is pulled away from the original house, creating a private courtyard space that transforms the layout and character of the original property. Central to the ground-floor layout, the courtyard unifies the building socially by establishing view corridors that interconnect living and garden spaces, and tectonically, by expressing the coming together of original and new elements.

The new rear wing on the ground floor encloses on the western boundary to form a courtyard intermediary between the original property and the new extended living area. An existing building footprint and a mature mountain ash tree generated the form of this zinc-clad wing. The extension leans away from the tree's canopy, pitching in height towards the house to create a vaulted ceiling over the new dining and living area, and lowering towards the courtyard to preserve sunlight access to the existing ground-floor rooms and garden outlook from the first-floor bedroom window. Openings in the existing external wall were maintained and provide access points to the new extension.

The original external fabric of the building was preserved and integrated into the extension. What were side windows looking directly to a boundary fence are now portals that allow circulation between the kitchen and new living area. Wide glass sliding doors frame new view lines established to the full length of the property. Whilst sequence has been preserved, these spaces are now interlinked by views to the courtyard, living wing and garden beyond, improving the daylight and vista quality of the interior without augmenting its spatial typology.

PROJECT SPECIFICATIONS

Location: North London, England
Area: 181 m² (1948 ft²)
Completed: 2015
Photography: David Butler

Inside the home, the soft grey hues and timber features of the original rooms contrast with the white walls and black-framed metal windows of the new additions. Externally, the graphite-zinc cladding defines new elements of the existing slate and brick. The subtleties of these differences are a result of meaningful interventions, which serve to define the property's history whilst giving it a unified elegance.

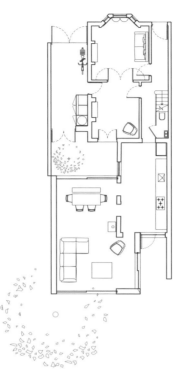

Ground-floor plan

First-floor plan

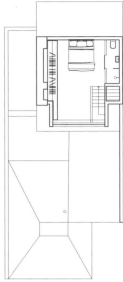

Second-floor plan

佩妮克罗夫特别墅

内皮尔·克拉克建筑事务所

Pennycroft is a five-bedroom family home that replaces an outdated 1930s house on a semi-rural plot. To comply with the area's strict planning policy the design is a contemporary interpretation of the local Arts and Crafts architecture. It has been positioned to occupy the front of the plot and align with adjacent properties.

Napier Clarke Architects

PENNYCROFT

The concept is simple – a brick base, punctured with a series of apertures supports clerestory windows with timber detailing above. Mirrored on both sides of the home, glass gables rise to an apex set within a high-pitched roof. Locally sourced, handmade brick with lime mortar softens the contemporary appearance.

The new build comprises two separate structural elements: a large two-storey building and a vaulted single-storey living space. These are linked via a glass walkway, which allows the volumes to read separately whilst physically joining to create an L-shaped footprint orientated to frame the rear garden. Two perpendicular axes inform the internal layout of the building and create a strong visual and physical connection from one end of the home to the other. From these axes, a sequence of single-volume spaces tuck behind partitioning walls, creating a balance between open and enclosed space.

The main entrance reveals a double-height hallway with a high timber-clad vaulted ceiling. Upstairs, a gallery landing bridges this space and physically separates family bedrooms from the guest suite, providing each side with privacy. Large expanses of glazing throughout the home flood spaces with an abundance of natural light.

Adopting a sustainable approach was a key aspect of the home's design. The crushed aggregate from the existing house was used as a sub base for the new project. To resolve a two-meter slope on site, the building has been excavated into the ground to minimise visual disturbance to adjacent properties. Low U-values have been achieved through optimum provision of insulation on external walls, therefore reducing the need for artificial heating. High G-value, double-glazing works in combination with the deep timber reveals on the upper storey, providing subtle solar shading to the south-facing façade and allowing for large expanses of glass.

Locally sourced Bovingdon red brick with lime mortar wraps around the entirety of the lower floor and all timber is FSC-certified Douglas fir.

This project successfully integrates into its sensitive surroundings and provides the client with a distinctive new home, one that meets the requirements of modern living.

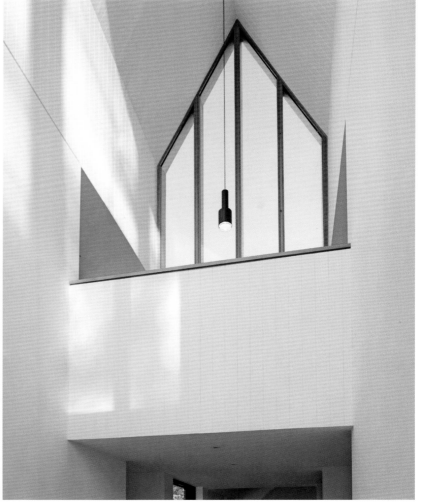

PROJECT SPECIFICATIONS

Location: Great Missenden, Buckinghamshire, England
Area: 430 m² (4628 ft²)
Completed: 2016
Photography: Joakim Borén

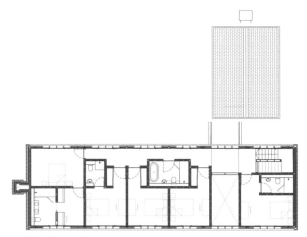
First-floor plan

Ground-floor plan

Ström Architects

THE QUEST

探寻别墅

电流建筑事务所

This new-build home replaced an ageing bungalow, which had been on the site since 1917. The clients wanted a house for retirement; one that demonstrated their keen interest in design and love of abstract and modern art but eschewed the usual trappings of staid retirement home design.

Early in the design process it was established that a single-storey building would suit the retired clients' future needs and the needs of their disabled daughter. An elegant solution to the steeply sloped site allowed for simple arrangements of space. The design also reduced visual impact from across the valley when looking back towards the house. The sloping site and its protected mature trees very much dictated the positioning of the dwelling.

The architects made use of a retaining wall faced in local Purbeck stone to define different levels and visually mask the vehicular route. This maintains a clear view from the living spaces across the valley. The house cantilevers over the retaining wall and creates a sheltered undercroft parking area. The large cantilever has been achieved through two concrete planes: the floor and roof acting together like a space beam. The concrete structure has been in-filled with simple timber dry lining, leaving the concrete frame externally visible.

Internally, a large open-plan kitchen/dining/living area occupies the centre of the plan, with a covered terrace spanning its length and providing a sheltered outdoor area overlooking southern views. To the Eastern end of the house are the guest accommodation and studies, with the master suite located within the cantilevered western end of the house.

A simple interior palette of Purbeck stone flooring, a concrete hearth and engineered timber flooring provide echoes of the external materials – waterproof concrete slabs and untreated larch cladding.

The house exemplifies the belief that simplicity and honesty of construction underpin good design.

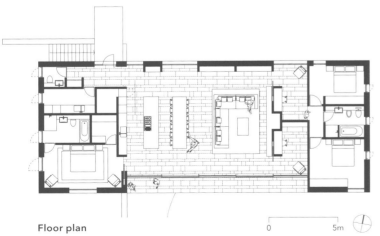

Floor plan 0 5m

PROJECT SPECIFICATIONS

Location: Swanage, Dorset, England
Area: 205 m² (2207 ft²)
Completed: 2015
Photography: Martin Gardner (pages 199, 200, 201); Michael Sinclair (pages 202, 203)

玫瑰园住宅

特里亚那·斯塔克建筑事务所

Rose Garden House borders an architectural conservation area on a granite plateau at the foot of the Dublin Mountains. The home's exposed timber beams and sedum roof allow it to blend with the landscape and sit quietly on the site.

The clients wanted a home that would reflect their personalities and lifestyles whilst feeling rurally idyllic despite its city location. Their extended family live either side of the site and requested that the home blend in with the landscape as much as possible. As a result, the home can barely be seen from the road and is deceptively private.

Triona Stack Architects

ROSE GARDEN HOUSE

The central spinal rooflight brings light and air to the heart of the house, whilst careful passive design maximises solar heat gain. The home's layout tracks the path of the sun. The bedrooms face east to the dawn light, the open-plan kitchen space opens to the southerly gardens, and the living room looks to the setting sun.

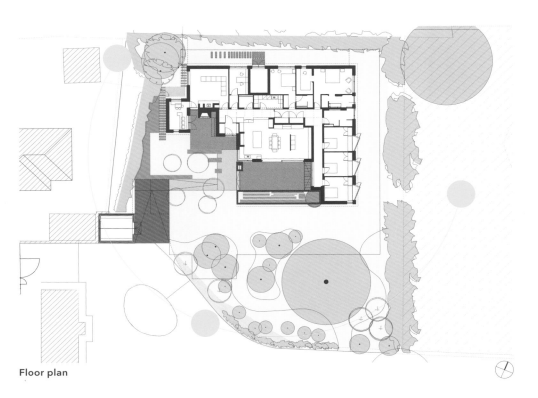

Floor plan

Large glass doors open out onto the patio and garden, creating a continuation of play space for the children and a haven of nature for the whole family to explore and enjoy. The south-facing aspect brings light and warmth into the main-living space throughout the seasons. Large brick walls bookend the space and huge wooden beams run across the room, providing great character and age that new builds can sometimes lack.

Built especially for a family with young children, the house allows for both shared times and quiet moments. It has been designed for lifetime living.

PROJECT SPECIFICATIONS

Location: Dublin, Ireland
Area: 300 m² (3229 ft²)
Completed: 2014
Photography: Alice Clancy

209

场景住宅

场景建筑事务所

Scenario House was purchased with the intention of completely renovating, extending and adapting to the scenarios of the architects' own family life. It presented them with an opportunity to practice their design principles on a more personal level.

The main design challenge was finding a way of physically and visually connecting the front part of the house, originally two separate Victorian reception rooms, with the basement level, which was a full floor height below.

Scenario Architecture

SCENARIO HOUSE

Creating an open, split-level, double reception – connected to the kitchen and garden floor by an angled, glazed-roof extension and a floating library feature leading up the bedroom floors – solved this challenge and created a sense of singular and connected space.

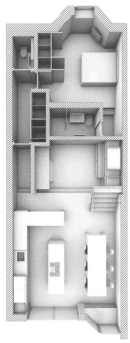 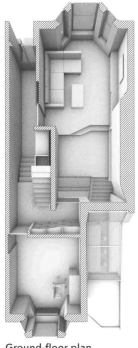 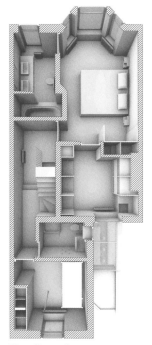 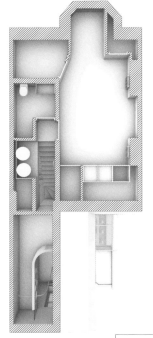

Lower ground-floor plan   Ground-floor plan   First-floor plan   Loft floor plan   0   5m

The design brief was created with consideration for both current and future scenarios, and a conscious effort was made to use every available space the house had to offer. This meant using spaces under the stairs, in the eaves, in hallways, and limited-height areas.

One such space included the relatively low-ceiling height area caused by the lowering of the reception above. This presented an opportunity for a fun and practical children's area. A storage unit under the stairs, complete with a pull-out table and bench, invites the children to spend time playing or doing homework and artwork in a designated but connected part of the home.

PROJECT SPECIFICATIONS
Location: London, England
Area: 220 m² (2368 ft²)
Completed: 2016
Photography: Matt Clayton

The architects' children played a central role in developing both the brief and the design of their own bedroom, which resulted in a climbing wall leading to a secret space at the eaves of the single-pitch original house extension and a fireman's pole.

斯普林菲尔德农场
安德鲁·伍德公司和Designscape建筑事务所

A contemporary and innovative two-storey house replaces what was an undistinguished 1950s bungalow. The site lies within the Bath and Bristol green belt on the edge of a scattered settlement in Stoney Littleton.

The building sits more or less on the footprint of the earlier building. It incorporates living accommodation on the ground floor and two en suite bedrooms on the first floor. The open-plan kitchen and dining area features full-height glazing, which overlooks the terrace to the

Andrew Wood Associates and Designscape Architects
SPRINGFIELD FARM

fields and views beyond. Solar gain is carefully controlled with heat mirror glazing technology and external blinds, which allow for comfortable room temperatures throughout the year and minimal heating costs. Multiple floor levels break up the living space and create a sense of zone variation within the plan. To the east of the building a book-lined snug is tucked away to one side. To the west lies a single-storey studio filled with light from full-height glazing as well as a bathroom and utility room. Adjacent to the house and its garden is a wild flower meadow.

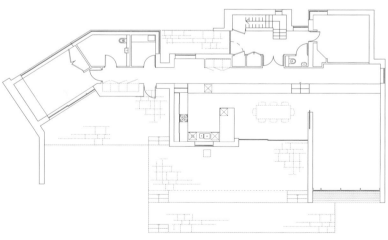

Ground-floor plan

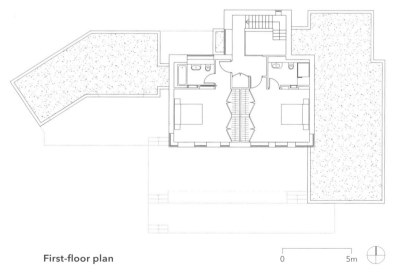

First-floor plan

The design of the building attempts to keep the house as low as possible on the sloping site, whilst simultaneously taking advantage of the attractive views across the Wellow Brook valley.

Natural and sustainable materials have been selected to complement the rural surroundings and colours of buildings in the area. The home's timber frame supports Frake hardwood cladding, which turns silvery grey as it ages and weathers, complementing the grey colours of local Lias limestone. Trespa Mateon cladding panels have been used on the stair tower and within the recesses to express plan articulation, whilst a pale grey Parex render has been used for the studio walls.

The aluminium window frames are a satin finish and coloured to blend with and complement the cladding and grey render. An extensive green roof on the studio is linked to the wild flower meadow by a raised bank. The banks are planted to blend with the adjacent meadow.

PROJECT SPECIFICATIONS

Location: Stoney Littleton, Bath, England
Area: 259 m² (2788 ft²)
Completed: 2014
Photography: Mark Bolton

圣约翰-赖伊住宅

玛尔塔·诺威卡

This project converted and redesigned a 1950s St John Ambulance station into a four-bedroom family retreat and creative workspace within the ancient citadel of Rye, in East Sussex.

The former St John Ambulance station was located within a private walled courtyard, which it shared with a Grade II–listed Georgian house.

Marta Nowicka – DOM stay and live
ST JOHN – RYE

After purchasing the property in 2013, the interior architect was challenged to create a space that would suit the aspirant lifestyle of Rye whilst also capturing its medieval history and the station's industrial past. By keeping the original lofty ground-floor area as living and working space and creating cosy bedrooms upstairs accessed via a large hallway, the brief was successfully implemented.

Removing a central wall and returning the station to its original floor plan created a large 72.5-square-metre (780-square-foot) open-plan living space on the ground floor. In keeping with the historical surroundings, the scheme of the design evokes medieval nuances with exposed-roof timber purlins and a central hearth from which the living space emanates. A double-sided wood burner sits within the vast entertaining space surrounding the hearth. A large concrete plinth cast at the centre divides the space into different zones. This medieval tone has been mixed with medical references, such as a stainless-steel nurses' trolley. The trolley has been used in the upstairs bathroom as a washstand/vanity unit with a large basin and mirror. The original ambulance doors were remade into an art piece and now feature in the living space.

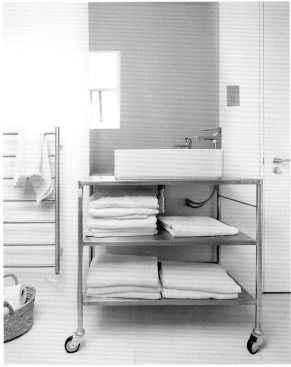

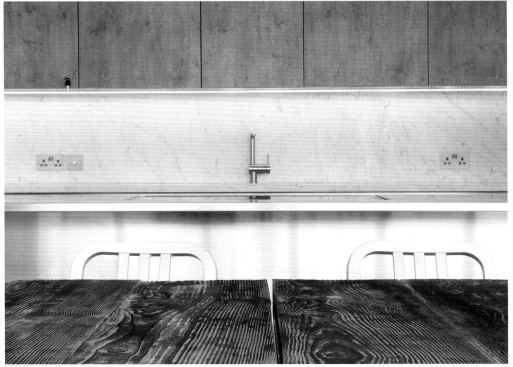

The overall style of the home is minimalistic with a mixture of textured materials, which add warmth. Original exposed brickwork and wide engineered oak flooring runs throughout the ground and first floors and continues to clad the walls of the double-height stairwell and bedrooms in the roof space. In the living space, a wrought ironwork light-fitting feature hangs over a large dining table made of reclaimed beach timber rescued from Camber Sands. The brushed stainless-steel kitchen island references the medical industry and complements the sandblasted, greyed-timber cabinets and stunning Carrera marble worktops and splashbacks.

The 1970s side extension was rebuilt. Drawing from modern local vernacular architectural inspirations, the new clad exterior of the extension – constructed with local Lydd handmade clay-peg tiles – fuses well with the existing roof-peg tiles and red-brick building, creating a unified and site-specific look. The surrounding listed garden brick walls influenced the door and window positioning. Bespoke architectural glass windows provide ample natural light and breathtaking rooftop and countryside views.

PROJECT SPECIFICATIONS

Location: Rye, East Sussex, England
Area: 215 m² (2314 ft²)
Completed: 2015
Photography: Voytek Ketz/Vision and Photography

First-floor plan

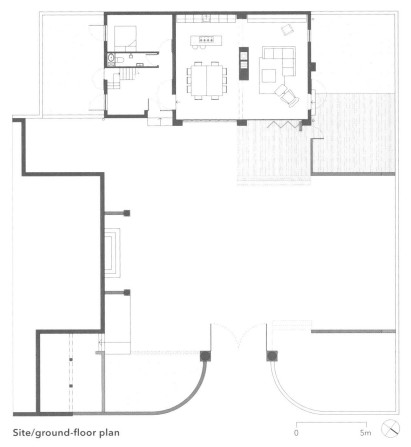

Site/ground-floor plan

茅草小屋

re|form 建筑与城市规划公司

Prior to the owners purchasing the house, the original cottage had been extended with a traditional brick and thatch component. The old extension created space in the new project for an updated kitchen and cloakroom facility, with bathroom and bedroom/study above.

re|form [architects + urban strategists]

THE THATCHED COTTAGE

The previous renovation, whilst providing nicely proportioned internal spaces, had failed to connect the house with the garden, so a more open and inclusive solution was sought. The aim of the project was to create a completely new space whilst utilising the massing of the previous extension and honouring the building's historic nature.

The bold move was taken to relocate the kitchen from its previous location into the new extension and free up the previous kitchen space to provide for a snug, which was better suited to an enclosed area of the house. This also achieved the desired link between the kitchen and external environment, allowing the client to feel connected to the garden even on days of inclement weather.

The design exploits the change of level across the garden to its advantage, recessing the room into the ground when viewed from the approach, which helps to reduce its impact upon the original property.

PROJECT SPECIFICATIONS

Location: Cheshire, England
Area: 138 m² (1485 ft²)
Completed: 2015
Photography: Kitchen Architecture

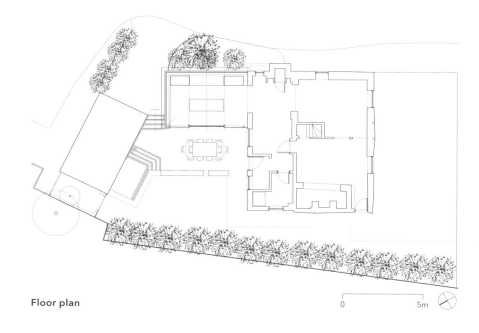

Floor plan

Materials and detailing were chosen to complement the existing house. The stone-flagged flooring provides continuity between the inside and the outside spaces, and the glazed walls of the new space are chased in to the existing brickwork, allowing the older structure to be read visually behind the new. Structural framing was kept to a minimum and used not only as a means of support to the glazed panels but also as a means of distributing services, power and lighting. The Bulthaup kitchen complements the simplicity of the space and was chosen for its bold form and minimal detailing. The client's own collection of classic furniture pieces add a touch of personal occupancy and warmth to the finished home.

泰瑞特·卡恩别墅

布莱德斯通建筑事务所

A rich cultural and natural environment influenced the design of this family dwelling in rural County Derry/Londonderry. Located on the periphery of Park Village, the land is surrounded by panoramic views of the Sperrin Mountains and is adjacent to a neolithic wedge tomb. Located further along Tireighter Road is a clachan displaying traditional characteristics of rural settlement, such as limewashed gables and boundary walls.

Broadstone Architects

TIREIGHTER CAIRN

Three distinct but inter-connecting blocks make up the home's form. The blocks were designed to convey a nearby burial chamber entrance and its three primary standing stones.

Similar to the clachan, each building bears a relationship with the other through proportional increases in scale, whilst clustered buildings step down the site via central circulation.

In this regard, the home replicates a local Palladian styled dwelling, in particular, Tamnagh Lodge, which prompted the design reference to Palladian villas and how a centrally lit entrance hall and circulation might work in accessing bedrooms and living/kitchen areas on upper, middle and lower floor plates.

Two of the three central entrance stacks act as both internal and external lightwells, whilst the third is a chimney for the living room stove. The two stairwells within the connecting spaces between the blocks allow for vistas from the entrance hall to the surrounding landscape and mountain range.

The materiality of the home makes reference to the wedge tomb's dense massing, the thick white gable walls of the clachan and the tall chimneys of Tamnagh Lodge. Building orientation, similar to the wedge-shaped tomb, provides a blank north façade with extending walls for privacy.

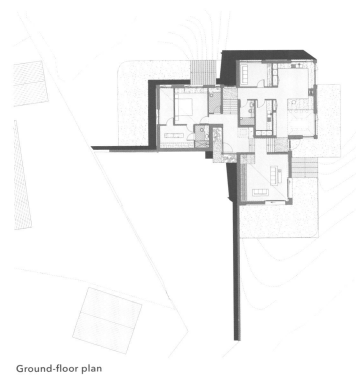
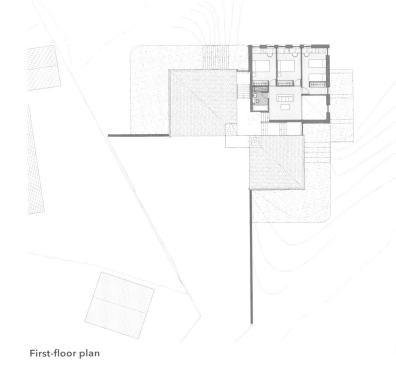

Ground-floor plan

First-floor plan

PROJECT SPECIFICATIONS
Location: County Derry/Londonderry, Northern Ireland
Area: 250 m² (2691 ft²)
Completed: 2016
Photography: Aidan Monaghan

温克利作坊住宅
柯克伍德·麦卡锡

Deteriorated and out of use for over 40 years, the existing single-storey workshop was demolished to expose an angled site. Surrounded on three sides by adjoining buildings and within a tightly held conservation area, the architects had to overcome numerous site challenges associated with density, rights to light, and planning stipulations on height. The result is a two-bedroom terrace built over three storeys.

Kirkwood McCarthy

WINKLEY WORKSHOP

Though humble to the street, the project is a voluminous residence referential to its setting yet of its own time. Stacked floor plates establish a natural zoning diagram within the property that transitions from the public/living basement floor to the private/sleeping first floor. Existing brickwork features, folded metal stairs and softened walnut floors articulate this program.

The home is entered via the street onto a split-level landing, which connects the basement with an upper-ground floor. This landing level, with perforated folded-steel stair, allows light from the southern street window to fall deep into the basement. This casts a gentle circular pattern on the brick basement floor below and forges a visual connection between the entrance and living area.

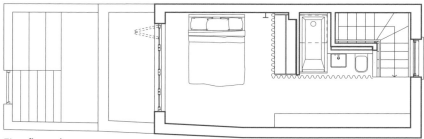
First-floor plan

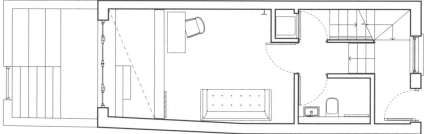
Ground-floor plan

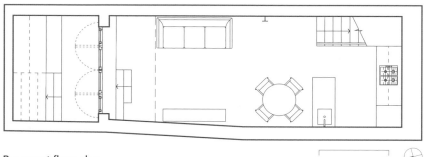
Basement floor plan

Open-plan living, dining and kitchen areas lead out onto the courtyard space at basement level. Soaring double-height glazing maximises internal daylight levels and sky views. It also integrates the living and outdoors spaces. The basement floor is covered with a herringbone-laid brick paver, which continues through to the courtyard, making the spaces feel like one shared indoor/outdoor experience. The basement-floor level steps up to the courtyard, and up again to the ground level of Teesdale Yard at the rear of the site.

PROJECT SPECIFICATIONS

Location: London, England
Area: 68 m² (732 ft²)
Completed: 2014
Photography: David Butler

At the upper-ground floor a flexible mezzanine bedroom/study space is divided by a bookcase and curtain from the living area and courtyard below. A private master bedroom with en suite and balcony occupies the top floor.

Externally, bricks localise the building within its setting. Internally, the palette is simple and continuous. The basement and courtyard are paved with herringbone bricks, which forge continuity between indoor and outdoor living spaces. A walnut stair with built-in cupboards leads from the basement to the ground floor and walnut flooring continues to the upper two storeys to distinguish bedroom spaces. Black folded-steel stairs punctuate circulation between the ground and first floors. The ground-floor metal stair is perforated to allow a passage of sunlight to enter the basement level, and the first-floor metal stair is enclosed with a suspended stud wall to provide an acoustic barrier to the master bedroom. The rest of the house has a simple and controlled material palette: white walls exaggerate light and space; black detailing highlights the journey through the house; and curtains for flexible privacy provide a softening element.

Ström Architects

WOODPECKERS

啄木鸟别墅

电流建筑事务所

Located in New Forest National Park, this private dwelling was designed as a two-storey replacement for a 1930s bungalow. The site sits amongst fields and woodland just outside the park, with a heavily treed and well-established garden.

The clients wanted a comfortable and contemporary home as a weekend house, one that would eventually become their future permanent home. With busy lives in London, they needed a weekend retreat to relax and entertain friends. As keen gardeners and outdoor lifestyle enthusiasts, a strong relationship with the garden as well as the surrounding countryside was an important design factor. Both involved in the design world, their aspirations for built and aesthetic qualities were high; they wanted a house that would be simple but elegant – stylish rather than fashionable.

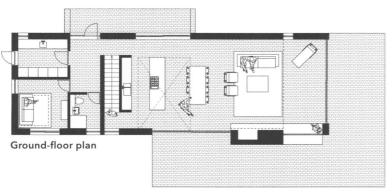

Ground-floor plan

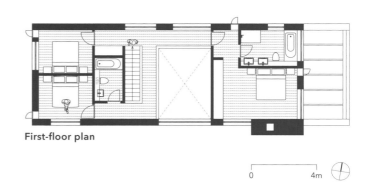

First-floor plan

0   4m

To create a strong internal-external relationship, traditional masonry to create large openings would be required, ultimately attracting costs and diluting the simplicity of a singular structural strategy. As it stands, the structure is entirely timber studwork and sits on a piled concrete slab. The studwork is filled and over-clad with insulation, creating a thermally and structurally efficient envelope. Planning constraints dictated that floor area could only be increased by 30 percent, with an additional 20 square metres (215 square feet) for a conservatory. Due to these restrictions it was established that a slab-sided volume would be best suited, pushing glazing to the outer edge of the envelope. By integrating a single-storey, almost flat-roofed element, which continues out seamlessly from the main open-plan space, the architects were able to avoid the typical aesthetic of a domestic conservatory.

The clients have ended up spending more time at the house than initially expected – long weekends, weeks spent working from home and extended stays hosting family during holidays – making Woodpeckers much more than a weekend getaway.

PROJECT SPECIFICATIONS

Location: New Forest National Park, England
Area: 195 m² (2099 ft²)
Completed: 2015
Photography: Luke Hayes

建筑事务所索引
INDEX OF ARCHITECTS

2020 Architects
www.2020architects.co.uk
Ballymagarry Road House 42–7

3S Architects and Designers
www.3s-ad.com
Lion Gate Gardens 146–51

Adam Knibb Architects
www.adamknibbarchitects.com
Austen House 28–33
Beckett House 48–53
Hurdle House 122–7

Andrew Wood Associates
www.andrewwoodassociates.co.uk
Springfield Farm 216–21

Ashton Porter Architects
www.ashtonporter.com
Drag and Drop House 66–71

Broadstone Architects
www.broadstone.ie
Oaklands 166–71
Tireighter Cairn 236–41

Bureau de Change
www.b-de-c.com
Folds House 78–83

Cassion Castle Architects
www.cassioncastle.com
Oak Lane House 158–65

Chiles Evans + Care Architects
www.cecastudio.co.uk
Artemis Barn 22–7
Knowle House 134–9

Coppin Dockray Architects
www.coppindockray.co.uk
Ansty Plum House + Studio 16–21

Dallas-Pierce-Quintero
www.d-p-q.uk
Courtyard House 60–5

Designscape Architects
www.dscape.co.uk
Springfield Farm 216–21

Dublin Design Studio
www.dublindesignstudio.com
Hazel Lane Mews 98-103

Gresford Architects
www.gresfordarchitects.co.uk
The Old Water Tower 172-7

Hudson Architects
www.hudsonarchitects.co.uk
Le Petit Fort 140-5

Jestico + Whiles
www.jesticowhiles.com
House 19 104-9

Kirkwood McCarthy
www.kirkwoodmccarthy.com
Pages Lane 184-9
Winkley Workshop 242-9

McCullough Mulvin Architects
www.mcculloughmulvin.com
One Up Two Down 178-83

Napier Clarke Architects
www.napierclarke.co.uk
Pennycroft 190-7

Neil Dusheiko Architects
www.neildusheiko.com
Brackenbury House 54-9
Gallery House 84-91

Marta Nowicka – DOM stay and live
www.martanowicka.com
St John – Rye 222-9

Platform 5 Architects
www.platform5architects.com
Backwater 34-41

re|form [architects + urban strategists]
www.reform-architects.london
The Thatched Cottage 230-5

SAM Architects
www.samarchitects.co.uk
MAP House 152-7

Sandy Rendel Architects
www.sandyrendel.com
142 South Street 10-15

Scenario Architecture
www.scenarioarchitecture.com
Scenario House 210-15

shedkm
www.shedkm.co.uk
House for an Art Lover 110-15

Spaced Out Ltd
www.spacedout.co.uk
Infinity House 128-33

A Small Studio
www.asmallstudio.co.uk
Escape to the Roof 72-7

Ström Architects
www.stromarchitects.com
Hartrow 92-7
The Quest 198-203
Woodpeckers 250-5

Triona Stack Architects
www.tsarchitects.ie
Rose Garden House 204-9

Tsuruta Architects
www.tsurutaarchitects.com
House of Trace 116-21

Every effort has been made to trace the original source of copyright material contained in this book. The publishers would be pleased to hear from copyright holders to rectify any errors or omissions.

The information and illustrations in this publication have been prepared and supplied by the contributors. While all reasonable efforts have been made to ensure accuracy, the publishers do not, under any circumstances, accept responsibility for errors, omissions and representations express or implied.